Pocket Guide to Rheology: A Concise Overview and Test Prep for Engineering Students

Xian Wen Ng

Pocket Guide to Rheology:
A Concise Overview
and Test Prep for Engineering
Students

 Springer

Xian Wen Ng
Singapore

ISBN 978-3-030-30584-0   ISBN 978-3-030-30585-7   (eBook)
https://doi.org/10.1007/978-3-030-30585-7

This Springer imprint is published by the registered company Springer Nature Switzerland AG
The registered company address is: Gewerbestrasse 11, 6330 Cham, Switzerland

Preface

Rheology is the study of flow and is therefore just as important and relevant to engineers as it is to other specialized professions in related fields such as the STEM disciplines.

This pocket guide serves as a supplementary learning resource that will help students deconstruct some of the most challenging problems encountered in the topic of rheology. There are up to 30 quality practice problems in this book, all of which are carefully selected to represent the most commonly encountered problems found in examination papers. With the comprehensive worked solutions and detailed explanations provided for each problem, students will be able to easily follow the thought process of problem-solving from start to finish, thereby honing their skills in applying abstract theoretical concepts to solve practical problems which is critical for acing examinations.

The balanced mix of both numerical and open-ended problems included in this book will also help students gain a well-rounded understanding of the topic of rheology, as they become proficient not only in handling numerical analysis but also in relating the significance of desktop problem-solving to the larger real-world context.

Singapore   X. W. Ng

Acknowledgments

My heartfelt gratitude goes to the team at Springer for their unrelenting support and professionalism throughout the publication process. Special thanks to Michael Luby, Aravind Kumar, Nicole Lowary, and Brian Halm for their kind effort and contributions toward making this publication possible. I am also deeply appreciative of the reviewers for my manuscript who had provided excellent feedback and numerous enlightening suggestions to help improve the book's contents.

Finally, I wish to thank my loved ones who have, as always, offered only patience and understanding throughout the process of making this book a reality.

Contents

About the Author

X. W. Ng graduated with First-Class Honors from the University of Cambridge, UK, with a Master's Degree in Chemical Engineering and Bachelor of Arts in 2011 and was subsequently conferred a Master of Arts in 2014. She was ranked second in her graduating class and was the recipient of a series of college scholarships including the Samuel Taylor Marshall Memorial Scholarship, Thomas Ireland Scholarship, and British Petroleum Prize in Chemical Engineering, for top performance in consecutive years of academic examinations. She was also one of the two students from Cambridge University selected for the Cambridge-Massachusetts Institute of Technology (MIT) exchange program in Chemical Engineering, which she completed with Honors with a cumulative GPA of 4.8 (5.0). During her time at MIT, she was also a part-time tutor for junior classes in engineering and pursued other disciplines including economics, real estate development, and finance at MIT and the John F. Kennedy School of Government, Harvard University. Upon graduation, she was elected by her college fellowship to the title of scholar, as a mark of her academic distinction.

Since graduation, she has been keenly involved in teaching across various academic levels, doing so both in schools and with smaller groups as a private tutor. Her topics of specialization range from secondary-level Mathematics, Physics, and Chemistry up to tertiary-level Mathematics and Engineering subjects. One of her recent publications include a practice book written for students taking engineering courses at the university. The book titled *Engineering Problems for Undergraduate Students* contains over 250 practice problems with worked solutions and stands as an all-in-one, multidisciplinary, problem-solving resource with comprehensive depth and breadth of coverage.

Fluid Types and Rheological Properties

Problem 1

Briefly explain what rheology means, and discuss why an understanding of rheological properties of materials (e.g., fluids) may be useful to engineers in the processing industries.

Comment also on some general methods adopted to study fluid flow of complex fluids, citing the example of viscoelastic fluids.

Solution 1

Worked Solution

What is Rheology?

Rheology is the study of a fluid's flow characteristics under an external force. Particularly, the deformation response of fluids under an applied stress (or strain) sheds light on interesting fluid properties under various operating conditions.

Complex fluids are widely prevalent in processing applications due to its primary characteristic in being able to flow and thereby doubling as a transport medium between unit operations in a typical processing plant. Fluids are relatively interesting as compared to solids because most fluids deform continuously in response to an applied force. Solids, on the other hand, tend to resist deformation and retain their shape until the applied force is sufficiently high to cause deformation, at which point they may instantaneously deform and fail (e.g., break apart) and deformation stops. Fully elastic solids like a metallic spring deform upon loading and return to their original shape (fully recoverable) when the load is removed. This is also called "elastic strain" as opposed to "plastic strain" when the material does not return fully to its original shape and is known to experience plastic deformation (not fully

© Springer Nature Switzerland AG 2019
X. W. Ng, *Pocket Guide to Rheology: A Concise Overview and Test Prep for Engineering Students*, https://doi.org/10.1007/978-3-030-30585-7_1

recoverable). It follows that compared to liquids, gases exhibit a greater deformation rate with their much lower molecular weights and weaker intermolecular forces.

Rheology plays a crucial role in almost all stages of material use and production and is therefore relevant in a wide range of industries. Understanding how rheologically complex fluids flow and respond to deformation forces helps engineers optimize product specifications and optimize processing conditions to achieve desired outcomes. Rheologically interesting fluids can be found in the formulation of many day-to-day household consumables from food products like mayonnaise, corn starch, and ketchup to cosmetics and shampoos. They also feature in larger-scale industrial applications, such as in the manufacture of paint, printer inks, car tires, and glues.

Below are some key considerations in rationalizing a fluid's response to applied deformation.

Intrinsic Factors

The molecular makeup of the material is an important factor. Some intrinsic characteristics include molecule shape and size, the presence of any charges on the particles, number of particles in a given volume, intermolecular separation distance, etc.

The molecular makeup of a fluid influences internal friction between fluid particles which each move at different velocities, thereby affecting viscosity and resultant flow patterns. In general, the larger the molecules (e.g., long-chain polymers or colloidal fluids), the greater the propensity to display complex behavior (non-Newtonian). On the contrary, most gases and fluids with low molecular weight exhibit simple behavior typically referred to as "Newtonian." Complex fluids may also display dynamic properties which vary over time and/or changes with operating conditions.

The type of interactions between molecules is also important. For example, whether molecules are bonded to each other by strong covalent bonds or relatively weaker hydrogen bonds will have a significant impact on rheological behavior.

External Factors

Fluids partially "resist" deformation forces through its viscosity, which is a measure of internal friction between fluid layers moving at different velocities (whereby higher viscosity leads to higher resistance to flow), and this resistance is akin to friction that has to be overcome in order to generate a flow.

External factors applied to a fluid directly affect its flow behavior since they alter flow conditions. A fluid can be subject to deformation forces that are aligned or unaligned. In the former case, aligned forces may pull the fluid apart leading to tension or press it together leading to compression, while, in the latter case, unaligned forces may result in bending, twisting, or shearing.

The applied deformation may also vary with time, such as in an oscillatory manner, or comprise of a combination of different deformation types.

Flow patterns can also be laminar or turbulent, where the latter introduces complexities due to irregular and rapid fluctuations in flow velocities. In channels and pipes, fluid flow can be driven by an applied pressure difference between opposite ends of the fluid passage, while, in reactor vessels, fluid motion can be driven by the rotation of an impeller. Other types of triggers for fluid flow include gravity which can cause a liquid film to fall downward and wall-driven flow whereby fluid may be dragged along by a moving solid surface at which the no-slip boundary condition applies.

Operating parameters in the surrounding environment also influence flow condition. Some examples include temperature, pH, solute concentration, and the presence of any additives such as polymers.

Methods to Study Fluid Flow

We can study fluids starting from a small fluid element and extrapolate this element to the entire fluid system to predict its bulk behavior.

Rheometers are typical instruments used to correlate applied forces and the fluid's deformation response. Experimental data obtained through measurements can then be fit to empirical or theoretical models, which provide a mathematical constitutive relationship between the input force (e.g., applied stress) and resultant deformation response (e.g., a strain response to an applied stress). This constitutive equation is typically expressed in the form as shown below, where τ denotes shear stress, η denotes viscosity, and $\dot{\gamma}$ denotes shear rate.

$$\tau = \eta\dot{\gamma}$$

When a fluid flows, a velocity gradient is created due to layers of fluid particles moving at different speeds. In simple pipe flow, friction between pipe walls and the flowing liquid contained within creates liquid layers where the layers closest to the walls move slowest. This velocity gradient between layers divided by their relative distance is in fact shear rate $\dot{\gamma}$. Shear stress τ, on the other hand, is defined as the force applied over a specified area. Therefore, the constitutive equation above relates the applied shear stress to the resultant shear rate, and the fluid's viscosity η acts to scale the magnitude of force necessary to shear a fluid to a specified degree.

Some common models used to predict the stress-strain relationship include the power law model, which is widely used for shear-thinning polymeric systems. Such rheological models can help us determine fluid viscosities, understand stresses present in the system, and hence determine useful quantities such as fluid flow behavioral limits, velocity fields, design limits for equipment, etc.

Viscoelastic Fluids

Many products we encounter in daily life, especially polymeric systems, belong to this category. These fluids lie between being perfectly viscous and perfectly elastic (like a metal spring). Under an applied force, a perfectly viscous fluid releases (or loses) all its energy via viscous dissipation and hence does not store any of it.

On the other hand, a perfectly elastic material stores all energy and loses none. We may sometimes model this type of elastic deformation as a linear Hookean spring where deformation length, x, is directly and linearly proportional to the external force, F, via a proportionality constant, k, as shown in the expression below.

$$F = kx$$

However, one should note that Hooke's law does not apply directly to most polymers as they tend to be non-linear elastics, unlike metals. They are often not perfect elastics and experience plastic deformation. The viscoelastic response is also time-dependent and correlates to the rate of applied stress. However, we can use linear models to study viscoelastic polymers if we limit to small magnitudes of strain such that deformation is still largely reversible like in a linear elastic.

When a polymer is stretched out, its chains are pulled apart and straighten. This reduces entropic freedom as the chains become more aligned and ordered, thereby creating a driving force for the polymer to restore its original state (i.e., unstretched state) which explains its elastic response. Most polymers between glass transition and melting point temperatures exhibit viscoelastic behavior. Above melting point, they behave like a viscous liquid, while, below the glass transition point, they behave like an elastic solid.

Types of Viscoelastic Responses

For viscoelastic fluids, there are a few characteristic responses to an external force.

1. Strain creep

When a constant stress is applied (e.g., step increase from zero to constant non-zero value at time $t = 0$) and resultant strain is measured, we observe that strain increases rapidly with time initially upon loading. After which, strain increases relatively more slowly with time until it plateaus at a maximum steady-state value. This time dependency of strain (or deformation) can be observed in the plot below as indicated in blue. We call this phenomenon "creep," and it is common in viscoelastics which experience increasing deformation under constant load. When the applied stress is released, the strain returns to its original value.

This is in contrast with perfectly elastic materials (below plot in orange) which do not deform any further no matter how long the load is applied. Strain instantaneously reaches a constant maximum value upon loading and holds that value under the constant applied stress.

Upon unloading, stress returns to zero instantaneously. On the other hand, for a perfectly viscous material (below plot in purple), the strain growth will be linear until removal of applied stress, at which point the strain value remains constant at that end value.

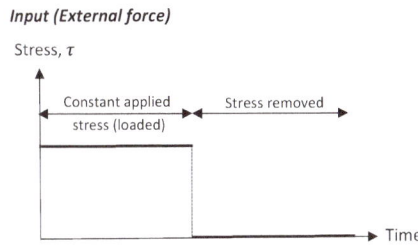

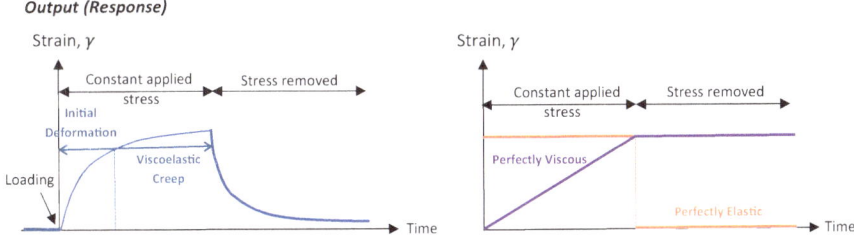

Real-Life Applications of Creep

There are several real-life examples of creep. We may observe creep even in the human body, such as in the bone which displays viscoelastic properties under a sustained load. Creep deformation occurs when there is an applied load over time, and this can cause damage to bone. A type of creep damage found in elderly individuals occurs in the vertebral bone of the spine and leads to signs of poor posture and decreased stature.

Another example of creep is in bone implants used for orthopedic procedures such as a total hip or knee joint replacement. A polymeric implant, typically an ultrahigh molecular weight polyethylene, is inserted into the bone to stabilize it or correct a misalignment. These implants may show a gradual reduction of thickness over time due to stresses that cause creep deformation, hence requiring revision surgery or replacement of worn implants with new ones.

In certain cases where implants are slightly malaligned at the start, shear stresses can develop at the bone-implant interface causing creep deformation in the bone over time, and this eventually leads to the loosening of implants from the bone-implant interface. Understanding the effects of creep in the abovementioned context can therefore help us design better materials for more durable implants, as well as improve upon surgical steps to achieve improved long-term outcomes.

Other than in the human body, we may also notice creep deformation in other day-to-day occurrences. Old glass windows are sometimes seen to be thicker at the bottom than at the top, and this is due to the downward-acting gravitational force causing strain creep in the glass. The glass panel deforms over time under this constant applied stress, resulting in the observed difference in thickness.

A Simple Experiment to Demonstrate Creep

We can demonstrate creep in a simple experiment using a rubber band which represents a viscoelastic material. As we subject the rubber band to a load (which represents a constant applied force or stress) hung at the bottom of the band, creep is observed as the rubber band strains and plastically deforms. As illustrated in the diagram below, the rubber band elongates vertically with time even when the applied stress is constant.

When unloaded, the band returns partially to its original shape, but there is some irreversibility in shape since it is not a perfect elastic. A notable characteristic about creep is that its effects are visible. In other words, when the rubber band is loaded, we can visually notice that the band continues to deform slowly and elongate over time, even though the load is constant. We will see later that in contrast, stress relaxation is not visible.

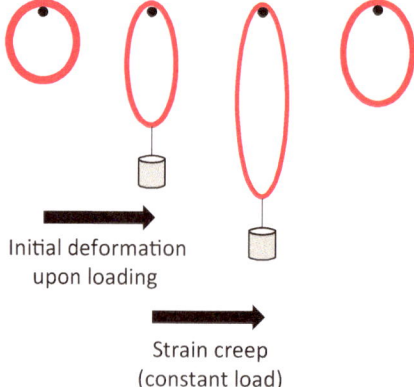

Initial deformation
upon loading

Strain creep
(constant load)

Dependency on Strain Rate

A viscoelastic material's creep pattern is dependent on strain rate. Some of us may notice that it is easier to break a piece of plastic when we pull it apart fast than if we stretched it out slowly. This means that at lower strain rates, the material is more deformable and able to elongate more without breaking. On the contrary, the material is less deformable at higher strain rates. This phenomenon has practical applications such as in examining biological tissue. The human tendon is an example of a viscoelastic material whereby the correlation between stress and strain is not linear like in a perfectly elastic material, but is instead non-linear due to a dependency of stress on strain rate (i.e., duration of load). As tendons are more deformable at lower strain rates, they absorb and store more mechanical energy which makes them less effective in carrying loads. At higher strain rates, tendons become stiffer and are less deformable; hence they are more effective in transmitting larger muscular loads to the bone as they absorb less energy in this state.

2. Stress Relaxation

Stress relaxation may be described as the gradual reduction of stress when the applied strain rate is zero. The applied strain rate can be zero when (i) strain itself is zero or (ii) when strain is non-zero but maintained at a constant value.

To demonstrate case (i), we may instantaneously unload a previously loaded material, such that the applied strain is removed. As for case (ii), we may artificially create a constant strain by holding the material in place with a fixed amount of deformation.

A Simple Experiment to Demonstrate Stress Relaxation

In a simple experiment using rubber bands, we can demonstrate stress relaxation for case (ii) described above. In this experiment, we apply a constant strain (or constant deformation) to the rubber band by stretching it to a given extended length and holding it there. The resultant stress developed in the material while the deformation is held constant is then measured over time. For a viscoelastic material, stress relaxation results in a reduction of stress over time in the loaded state even though strain (or deformation) is held constant.

Note here that the rubber band does not change shape during stress relaxation in the loaded state; hence stress relaxation has no visible effect unlike strain creep which is observable. Stress relaxation only shows visible effects when the load is removed, and we notice that the viscoelastic material only partially returns to its original shape. It is worthy to note that a perfectly elastic material will fully restore its original shape as it uses all the energy earlier stored during deformation to return to its original state (no energy is lost).

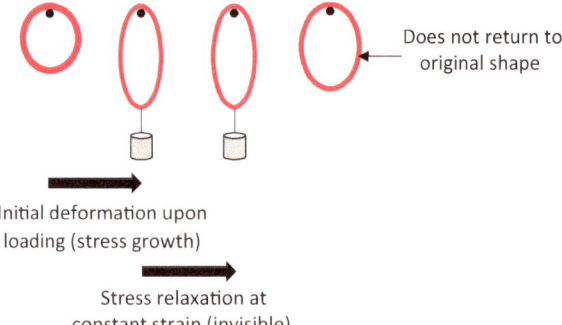

We can show the effects of stress relaxation graphically whereby the relaxation phase corresponds to the period when the applied strain rate is zero. Note that in our rubber band experiment, we have zero strain rate during stress relaxation since strain (or deformation) is held constant and does not vary with time (hence zero strain rate).

Viscoelastic Behavior

Output (Response) – Viscoelastic

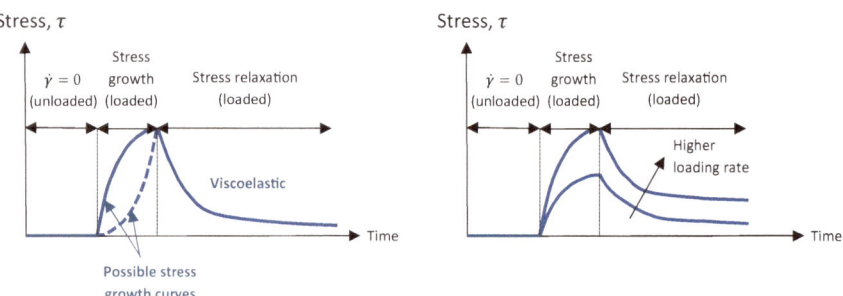

Stress relaxation arises from the molecular rearrangements that occur during deformation. When we pull a polymeric material such as a rubber band to a particular extended length, the polymer chains align to the axis of the applied force, and over time stress is reduced (hence relaxation).

The stress relaxation response is also dependent on strain rate, such that when the strain rate is higher (i.e., faster loading rate), the peak value of stress reached is higher. With a higher initial stress value at the start of the stress relaxation phase, the final value of stress reached is also lower. Conversely, at lower strain or loading rates, the material has more time to relax during loading, which leads to a lower peak stress reached. Consequently, stress relaxation phase leads to a lower final value of stress. This behavior is shown in the plot (above right).

For the purpose of analysis, we may also examine the viscoelastic response relative to the two extreme limits, i.e., a fully elastic and a fully viscous material. For a perfectly elastic material, stress does not reduce and instead maintains at its constant maximum value at all times. As for a perfectly viscous material, stress instantaneously vanishes to zero (even though strain is non-zero and held constant) when applied strain rate is zero due to complete dissipation of energy.

Fully Viscous Behavior

We can plot the stress response of a Newtonian fluid by applying the constitutive relationship, $\tau = \eta \dot{\gamma}$ whereby stress mirrors the shape of the strain rate plot, since fluid viscosity η is a constant.

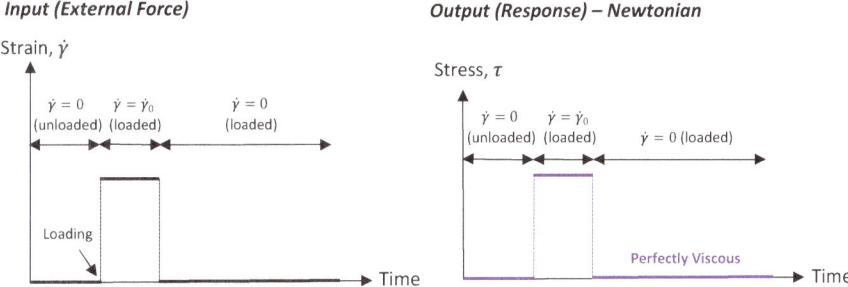

Fully Elastic Behavior

In examining fully elastic behavior, a useful plot to deduce from the strain rate profile is the strain function as shown below. This allows us to directly mirror the shape of the strain function for the expected stress profile for a perfectly elastic material since stress is directly proportional to strain with a proportionality constant that is often known as the elastic constant g in the equation $\tau = g\gamma$.

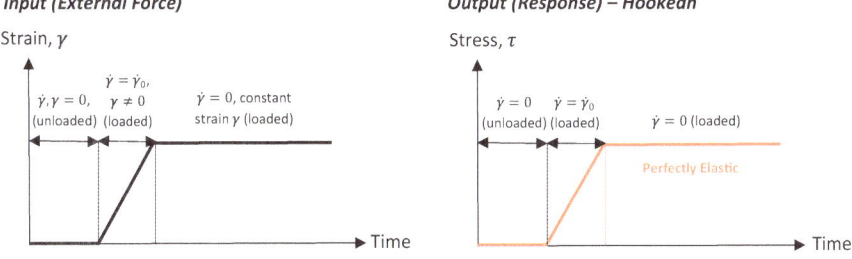

Applications of Stress Relaxation

This phenomenon can be observed in surgical procedures such as when a femoral nail implant is inserted into the femur bone. The resultant stress relaxes over time such that the nail can be pushed further into the bone at a later time, hence achieving the desired stabilizing effect of the implant without creating undue stress on the bone.

3. Oscillatory Applied Stress

This is similar to strain creep in that an applied stress produces a strain response. However, instead of a constant stress, the applied stress follows a sinusoidal function with time. For viscoelastic materials, this produces a similarly oscillatory strain response, whereby the strain lags behind stress by an amount known as phase angle or phase shift. The degree of lag is largely determined by material properties. This phase lag also partially explains why we observe hysteresis in viscoelastic materials whereby the loading curve does not follow the unloading curve in a stress-strain plot.

4. Hysteresis

Hysteresis can be explained by tracing where energy goes in the system. When a viscoelastic material is loaded, both strain and stress increase. Due to the material's partial elastic property, some of the energy earlier stored during loading is used to reverse the deformation upon unloading. On the other hand, the partial viscous property of the material causes some of the earlier stored energy to be dissipated and lost as heat energy (unrecovered) upon unloading. We can better understand hysteresis from the plots below.

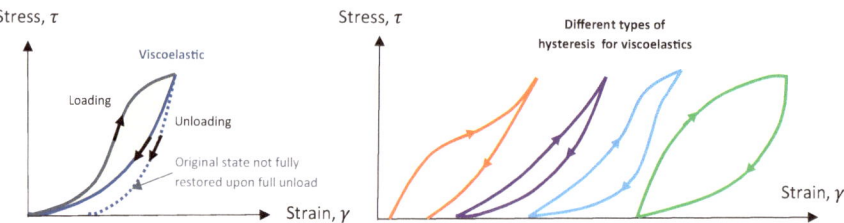

The area under the loading line represents energy stored during loading. The area under the unloading line represents energy returned to the system during unloading which is used to reverse the deformation. Therefore, the difference between the two areas, represented by the area between the two curves as shown in yellow above, is equivalent to energy that is lost. The degree of hysteresis is a measure of this difference and scales with molecular weight of the fluid.

Depending on the type of viscoelastic material, the loading and unloading curves can take various forms although all would be non-linear curves. It is worth noting that some materials may not restore their original shapes fully upon unloading, which means some degree of permanent deformation occurs. This is shown in dotted line in the plot above.

Fully Elastic Response

Let us examine the special case of a perfectly elastic material whereby stress is directly proportional to strain via a proportionality constant; therefore the stress-strain plots are straight lines as shown below. In addition, the loading and unloading lines are the same line as no energy is lost or dissipated. All the energy stored in the material during loading is used to reverse the deformation (i.e., fully restore to its original shape) during unloading. It follows then that the area between the loading and unloading plots will be zero, and this gives the same line as shown below in orange.

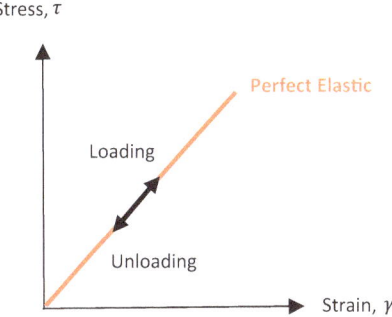

5. Viscosity

When discussing viscosity, it is useful to note two major dependencies:

- Dependency on time (with all other variables held constant, e.g., under constant shear over time)
- Dependency on the magnitude of the applied force (e.g., shear or stress)

Viscoelastic fluids are rheologically complex and so are their viscosity profiles. Before we go further into viscoelastic fluids, let us first recap the various categories of fluids so that we may appreciate where viscoelastic fluids fit in in the grand scheme of rheological types.

Note that real fluids differ from ideal fluids (hypothetical) in that the former has non-zero viscosity while the latter has zero viscosity. In the following discussion, we are dealing with real fluids with non-zero viscosity.

Newtonian fluid (Viscous, inelastic)		Non-Newtonian fluid (Viscous, inelastic)		Non-Newtonian fluid (Viscous, elastic)	Hookean solid (Inviscid, elastic)
$\tau = \eta\dot\gamma$, where η is a constant viscosity $\eta \neq f(\dot\gamma, t)$ Viscous (inelastic), therefore time-independent shear stress $\tau \neq f(t)$ Dissipates all energy, stores none		$\tau = \eta_{app}\dot\gamma$ (not perfect Newtonian) $\eta_{app} = f(\dot\gamma), \eta_{app} \neq f(t)$ Viscous (inelastic), therefore time-independent shear stress $\tau \neq f(t)$ Dissipates all energy, stores none		$\tau = \eta_{app}\dot\gamma$ (not perfect Newtonian) $\eta_{app} = f(\dot\gamma, t)$ Has viscous and elastic properties. Elasticity leads to time-dependent shear stress $\tau = f(t)$ Dissipates some energy, stores some energy within system **(h) Viscoelastic**	**$\tau = g\gamma$, where g is the elastic constant** Elastic (inviscid) Stores all energy within system, dissipates none **(i) Hookean solid**
No yield stress **(a) Newtonian liquid**	**Has yield stress** **(b) Viscoplastic (e.g., Bingham Newtonian liquid)**	**No yield stress** **(c) Pseudoplastic** **(d) Dilatant** **(e) Thixotropic** **(f) Rheopectic**	**Has yield stress** **(g) Viscoplastic (e.g., Bingham plastic)**		

(a) Newtonian Liquid

Before we delve into the viscosity characteristics of more complex fluids, let us first establish some properties of the simple Newtonian liquid. The viscosity of a Newtonian liquid does not change over time under a constant applied stress (force per unit area). The constant value of viscosity is determined by the corresponding values of $\dot{\gamma}$ and τ according to the equation below.

$$\tau = \eta\dot{\gamma}$$

Another way to describe the time-independent property of viscosity is to say that it responds "instantaneously" to applied shear, which means that if there were any changes to the magnitude of the constant applied shear stress, viscosity adjusts instantaneously to the corresponding new value. (Note that this time-independent characteristic is not exclusive to Newtonian liquids. Inelastic non-Newtonian fluids may also exhibit time-independent viscosities. This will be further discussed under pseudoplastics and dilatants).

In addition to the abovementioned time-independent property for Newtonian liquids, the viscosity of Newtonian liquids also does not vary with shear rate. This is the reason such liquids are called "Newtonian" since they obey Newton's law where η in the equation above is a constant.

A viscous liquid (inelastic) is often modelled using a dashpot as illustrated below, which is a damping device that resists motion via viscous friction from the fluid contained in the dashpot.

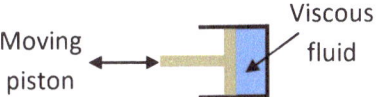

The piston of the dashpot moves slower with a more viscous fluid and faster with a less viscous fluid. This model is also useful in helping us understand that viscous fluids do not store any energy like elastic solids, but instead, all energy is lost as dissipated heat. This is also the reason that deformations are irreversible for viscous inelastic materials as they are not able to restore their original shape partially or fully without any stored energy in the system to achieve this restoration.

In summary, here are some key facts about Newtonian liquids:

Typical physical states (Examples)	Liquid, gas (e.g., water, mineral oil, air)
Constitutive equation	Follows Newton's law $\tau = \eta\dot{\gamma}$, where η is a constant
Shear stress τ	Scales linearly with shear rate
Viscosity	Viscous (real fluid)
Elasticity	Inelastic, no energy is stored as all dissipated, hence fluid does not return to original shape after applied stress is released
Viscosity function, η	Not dependent on time and shear rate $\eta \neq f(\dot{\gamma}, t)$

(b) Viscoplastic (e.g., Bingham Newtonian liquid)

The Bingham Newtonian liquid is similar to the Newtonian liquid, whereby the only difference is that it has a yield stress. This means that if the applied shear stress is below the yield stress value, the liquid will not deform (or flow for a liquid) and only flows above this yield value. This is unlike the Newtonian liquid which does not have such a yield stress; hence it deforms under any application of stress.

Mathematically, the constitutive equation is slightly adjusted to account for the yield stress τ_0. Once the fluid is set in motion (deforms when applied shear stress τ>yield stress τ_0), it behaves like a Newtonian liquid.

$$\tau = \tau_0 + \eta\dot{\gamma}$$

In summary, here are some key facts about Bingham Newtonian liquids:

Typical physical states (Examples)	Liquid (some slurries)
Constitutive equation	Follows Newton's law once yield stress is exceeded $\tau > \tau_0$, then $\tau = \eta\dot{\gamma}$, where η is a constant
Shear stress τ	Scales linearly with shear rate
Viscosity	Viscous (real fluid)
Elasticity	Inelastic, no energy is stored as all dissipated, hence fluid does not return to original shape after applied stress is released
Viscosity function, η	Not dependent on time and shear rate $\eta \neq f(\dot{\gamma}, t)$

(c) Pseudoplastic

Unlike the Newtonian fluids, pseudoplastic fluids are called non-Newtonian because their viscosities are no longer a constant denoted by η, but instead vary with shear rate $\dot{\gamma}$. It follows then that they do not obey Newton's law where η is a constant. Instead, we replace the constant η with an apparent viscosity η_{app} which is a function of shear rate $\dot{\gamma}$.

$$\tau = \eta_{app}\dot{\gamma}$$

Pseudoplastics are shear-thinning, and this means they exhibit a decreasing apparent viscosity with increasing shear rate. They are more common than shear-thickening fluids or dilatants.

It is worth highlighting that the viscosity of pseudoplastic fluids is time-independent and does not change over the duration of application of shear. This is assuming that other than time, all other variables including $\dot{\gamma}$ are held constant, under a constant applied stress (force per unit area). This distinction will become clearer when we introduce the category of thixotropic fluids. The constant value of viscosity is determined by the corresponding values of $\dot{\gamma}$ and τ according to the equation $\tau = \eta_{app}\dot{\gamma}$ where η_{app} is a function of shear rate $\dot{\gamma}$.

In summary, here are some key facts about pseudoplastic fluids:

Typical physical states (Examples)	Liquid (small particle suspensions, polymer solutions, polymer melts, printing inks, blood, mayonnaise, ketchup, quicksand, milk, paints, nail polish)
Constitutive equation	Does not follow Newton's law, instead follows $\tau = \eta_{\mathrm{app}}\dot{\gamma}$, where η_{app} is a function of $\dot{\gamma}$
Shear stress τ	Scales non-linearly with shear rate and decreases with increasing shear rate (shear-thinning). However, shear stress is not a function of time
Viscosity	Viscous (real fluid)
Elasticity	Inelastic, no energy is stored as all dissipated; hence fluid does not return to original shape after applied stress is released
Viscosity function, η	Not dependent on time but dependent on shear rate (decreases with increasing shear rate)

(d) Dilatant

Unlike Newtonian fluids, but similar to pseudoplastic fluids, dilatants are also non-Newtonian since their viscosities vary with shear rate $\dot{\gamma}$. Therefore, dilatants follow the expression below where apparent viscosity η_{app} is a function of shear rate.

$$\tau = \eta_{\mathrm{app}}\dot{\gamma}$$

However, unlike pseudoplastics, the apparent viscosity of dilatants increases with increasing shear rate. Dilatants are therefore called shear-thickening. Similar to Newtonian fluids and the pseudoplastic fluid, the viscosity of dilatants does not vary with time.

In summary, here are some key facts about dilatant fluids:

Typical physical states (Examples)	Liquid (large amount of ultrafine, irregular, deflocculated particles dispersed in a continuous phase in suspension, gum solution, aqueous suspensions of titanium dioxide, quicksand, starch suspensions)
Constitutive equation	Does not follow Newton's law, instead follows $\tau = \eta_{\mathrm{app}}\dot{\gamma}$, where η_{app} is a function of $\dot{\gamma}$
Shear stress τ	Scales non-linearly with shear rate and increases with increasing shear rate (shear-thickening). However, shear stress is not a function of time
Viscosity	Viscous (real fluid)
Elasticity	Inelastic, no energy is stored as all dissipated; hence fluid does not return to original shape after applied stress is released
Viscosity function, η	Not dependent on time but dependent on shear rate (increases with increasing shear rate)

(e) Thixotropic

Thixotropic fluids can be thought of as an extension from pseudoplastics, whereby they are time-dependent pseudoplastic fluids. Like pseudoplastics, thixotropic fluids are non-Newtonian and do not have a constant viscosity. Instead their viscosity varies with certain variables. Thixotropic fluids are similarly shear-thinning with a reducing viscosity with increasing shear rate.

However, other than having a viscosity that varies with shear rate, the additional property that thixotropic fluids have over pseudoplastic fluids is that their viscosity is also a function of the time over which force is applied (shear or stress). Thixotropic fluids show a reducing viscosity over time at a constant shear rate; hence the longer the fluid is subject to shear stress, the lower is its viscosity. Such fluids are thereby described as fluids that "take time" to achieve the final equilibrium viscosity when there is a step change in shear rate, and this is in contrast with the instantaneous response of viscosity to the corresponding applied shear rate for non-time-dependent fluids such as pseudoplastics, dilatants, and Newtonian fluids. In the same way, when the shear rate is decreased back to a lower value, viscosity returns to its original higher value but at a slower rate than pseudoplastics due to the time-dependent characteristic.

The difference between the downward and upward curves (shown below left) in thixotropic systems is called the degree of hysteresis. The larger the molecular weight of the fluid, the greater the degree of hysteresis since it makes sense that bigger molecules take longer to recover to full viscosity. Pseudoplastics require nearly zero time for recovery, giving an elastic solid-like response (no hysteresis). On the other hand, thixotropic fluids take time to re-establish itself when left at rest after shearing and require some recovery time to regain its higher viscosity, giving a response similar to that found in viscous or viscoelastic shear-thinning fluids where some degree of hysteresis is observed.

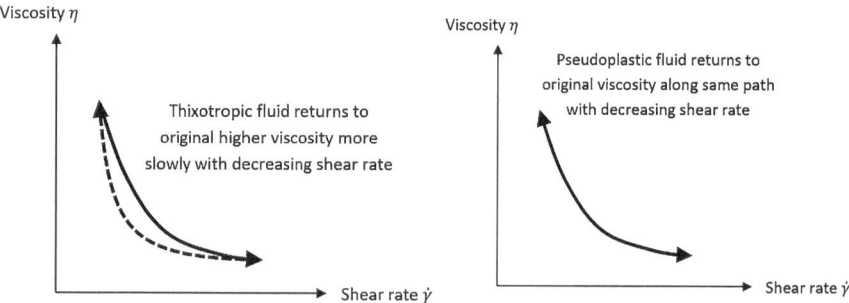

Another good way to explain the thixotropic (time-dependent) concept is shown below, where we see the time-dependent quality from the data that represent different periods of application of shear and the shear rate-dependent quality from the same type of curvature of the plots, regardless of the duration of application of shear. The gradient of the plot below is equivalent to apparent viscosity given by the constitutive expression $\tau = \eta_{app}\dot{\gamma}$.

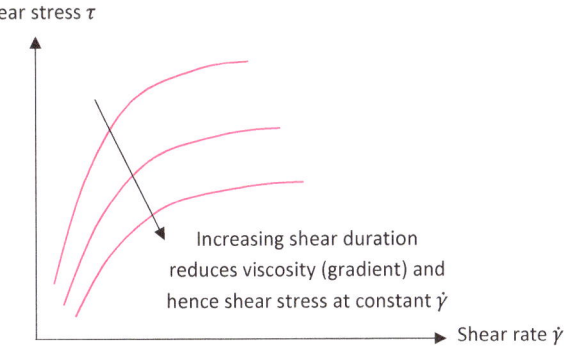

In summary, here are some key facts about thixotropic fluids:

Typical physical states (Examples)	Liquid (crude oil, grease, heavy printing inks, drilling fluid, honey, paint, ketchup, nail polish)
Constitutive equation	Does not follow Newton's law, instead follows $\tau = \eta_{app}\dot{\gamma}$, where η_{app} is a function of both $\dot{\gamma}$ and time t
Shear stress τ	Scales non-linearly with shear rate, decreasing with increasing shear rate (shear-thinning) Shear stress is also a function of time, decreasing with prolonged application of shear
Viscosity	Viscous (real fluid)
Elasticity	Inelastic, no energy is stored as all dissipated; hence fluid does not return to original shape after applied stress is released
Viscosity function, η	Dependent on both time and shear rate (decreases with increasing values of both)

(f) Rheopectic

Similar to thixotropic fluids, rheopectic fluids are an extension from dilatants in that their viscosities are also time-dependent in addition to being shear rate dependent and in a shear-thickening way. This means that rheopectic fluids exhibit an increasing viscosity with increasing shear rate as well as with increasing duration of shear stress.

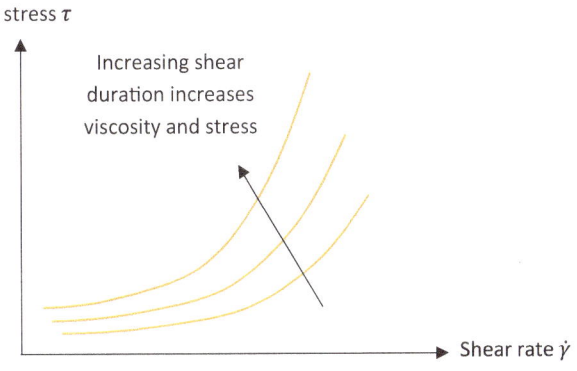

In summary, here are some key facts about rheopectic fluids:

Typical physical states (Examples)	Liquid (whipped cream, egg white)
Constitutive equation	Does not follow Newton's law, instead follows $\tau = \eta_{app}\dot{\gamma}$, where η_{app} is a function of both $\dot{\gamma}$ and time t
Shear stress τ	Scales non-linearly with shear rate, increasing with increasing shear rate (shear-thickening) Shear stress is also a function of time, increasing with prolonged application of shear
Viscosity	Viscous (real fluid)
Elasticity	Inelastic, no energy is stored as all dissipated; hence fluid does not return to original shape after applied stress is released
Viscosity function, η	Dependent on both time and shear rate (increases with increasing values of both)

(g) Viscoplastic (e.g., Bingham plastic)

This type of fluid is similar to the Bingham Newtonian liquid covered earlier, in that it possesses a yield stress, thereby requiring a minimum magnitude of applied shear stress before the fluid will be set in motion (or flow). A viscoplastic fluid however, unlike the Bingham Newtonian liquid, exhibits non-Newtonian properties of shear-thinning or shear-thickening. Therefore, the difference here between the two types of Bingham fluids is that the viscoplastic Bingham fluid has a viscosity that is a function of shear rate. However, similar to pseudoplastics and dilatants, their viscosity is not a function of time.

Mathematically, the constitutive equation is slightly adjusted to account for the yield stress τ_0 as well as the non-Newtonian apparent viscosity. Once the fluid begins to deform when $\tau > \tau_0$, it behaves like a pseudoplastic or dilatant fluid.

$$\tau = \tau_0 + \eta_{app}\dot{\gamma}$$

In summary, here are some key facts about viscoplastic Bingham fluids:

Typical physical states (Examples)	Liquid (clay suspensions, oil paint, drilling mud, toothpaste)
Constitutive equation	Does not follow Newton's law, instead follows $\tau = \eta_{app}\dot{\gamma}$, where η_{app} is a function of $\dot{\gamma}$
Shear stress τ	Scales non-linearly with shear rate, increasing with increasing shear rate (shear-thickening) or decreasing with increasing shear rate (shear-thinning) Shear stress is not a function of time
Viscosity	Viscous (real fluid)
Elasticity	Inelastic, no energy is stored as all dissipated; hence fluid does not return to original shape after applied stress is released
Viscosity function, η	Dependent on shear rate (increases or decreases with increasing shear rate)

(h) Viscoelastic

Viscoelastic fluids exhibit both viscous and elastic properties which vary in relative extents under different conditions. Therefore, their behavior lies between that of a fully viscous inelastic liquid and a fully elastic solid.

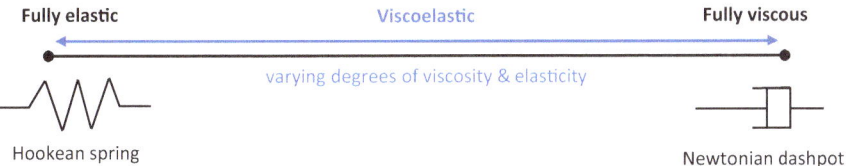

The specific mathematical functions that describe shear stress and apparent viscosity for such fluids are more complicated as there is a diverse range of possible combinations for the viscous and elastic parts of the fluid. To start, we may first appreciate that shear stress for a viscoelastic fluid will have some dependency on strain rate arising from the viscous part and some dependency on strain arising from the elastic part.

It is useful to understand some basic models (e.g., the Maxwell and Voigt models) used to fit viscoelastic behavior. These models are appropriate as they fulfill the basic requirement of being able to model viscosity and elasticity in the same system. They are however simplistic for several reasons, for example, they consider only linear behavior for both the viscous and elastic components and are thus inadequate for modelling non-linear shear-thinning. In addition, they consider only a single element for each property (i.e., single dashpot and single spring) which is seldom robust enough to represent real complex fluids in processing applications. More complicated combinations of these models are sometimes used to better fit with real viscoelastic fluids and their complex behavior.

Maxwell Model

The Maxwell model consists of a single dashpot connected to a single spring in *series*. A Maxwell element is illustrated as shown below.

The deformation profile for a Maxwell model is therefore a linear sum of the individual component's profiles.

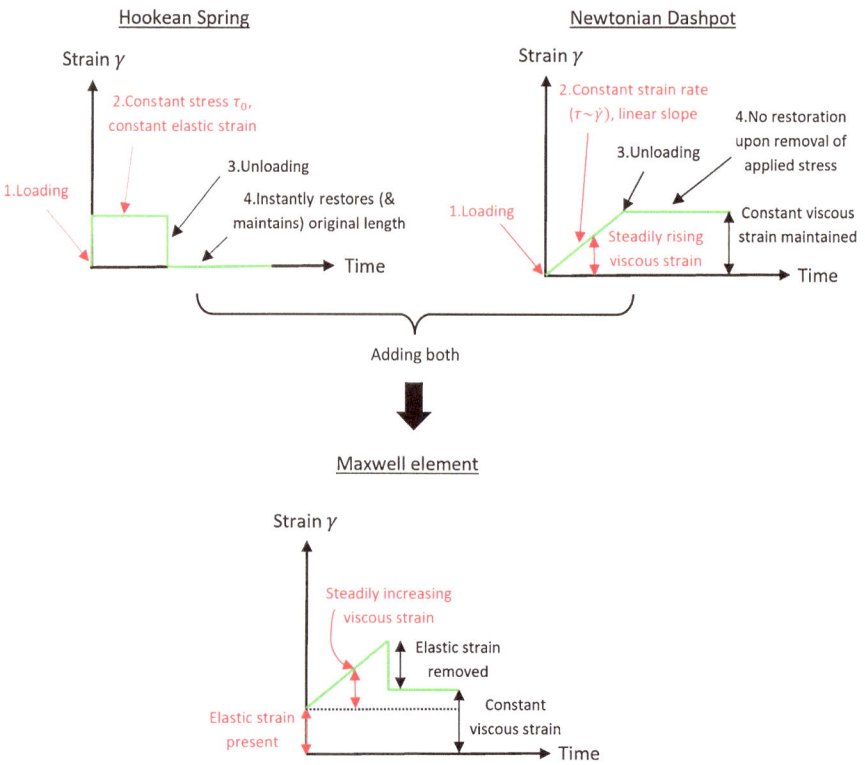

Key Features of the Maxwell Model

One of the key points to note about the Maxwell element is its inadequacy in modelling non-linear increase in strain for a viscoelastic fluid (including initial deformation and subsequent strain creep) under a constant applied stress. This is because the sum of both the spring and dashpot strain profiles gives a linear strain increase with time as illustrated in the above graphs. The sharp corners observed in the plot make the model unsuitable for describing creep. In addition, it shows a continually increasing strain (or deformation) for as long as stress is applied, unlike actual creep behavior which has an eventual "plateauing."

To address some of these limitations, the Maxwell model can be adjusted to better fit real data, e.g., by combining more than one Maxwell element together and/or by introducing additional parameters to the model such as damping factors.

The Maxwell element is however good at predicting strain response upon removal of applied stress, since the series connection (unlike a parallel arrangement such as in a Voigt model) makes it able to portray the characteristic whereby some of the deformation created during loading is never restored due to the viscous component of the viscoelastic fluid (energy lost as dissipated heat).

Voigt Model

The Voigt model consists of a *parallel* connection of a single dashpot and a single spring (also called parallel coupling).

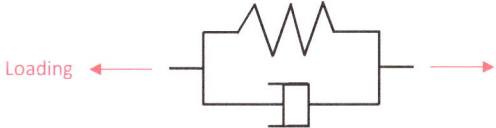

The deformation profile of a Voigt element is shown below.

Voigt element

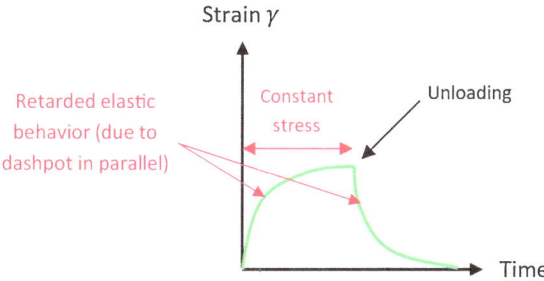

Key Features of the Voigt Model

A single Voigt model predicts non-linearity better than the Maxwell model for both creep and relaxation. The parallel connection means viscosity posed by the dashpot "forces" the magnitude of strain to rise slowly from zero instead of instantaneously. This works against the spring component's tendency to instantaneously deform to maximum strain upon application of stress.

Contrary to the Maxwell model, the Voigt model is less ideal for modelling any permanent deformations that might result from the viscous component of real viscoelastic fluids. This is because when applied stress is removed, the parallel connection with a fully elastic spring makes the initial deformation completely reversible (i.e., no energy lost through viscous dissipation). The Voigt model is also unable to demonstrate any instantaneous responses that might result from the elastic component of real fluids.

Combination of Models for Viscoelastic Fluids

The combination of models allows the viscoelastic properties of non-linear strain increase (during loading) and non-linear strain decrease (at partial recovery) to be portrayed. One of the ways to combine both the Maxwell and Voigt models for complementary benefits in modelling creep-recovery is to connect a Voigt model in series with another Maxwell element.

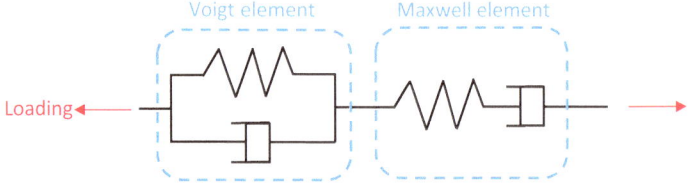

The deformation profiles for this connection are shown below.

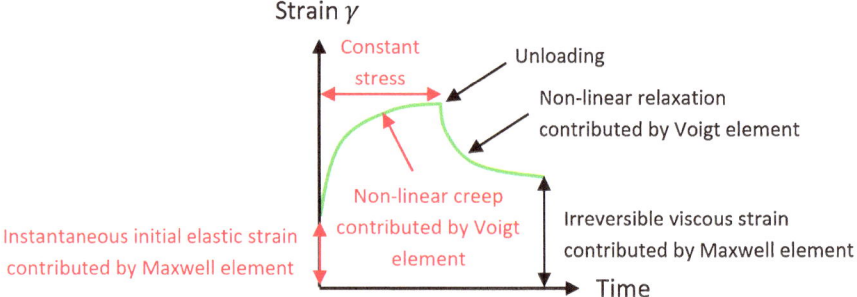

Another useful way to combine models is to connect a Maxwell element with a Newtonian dashpot as shown below.

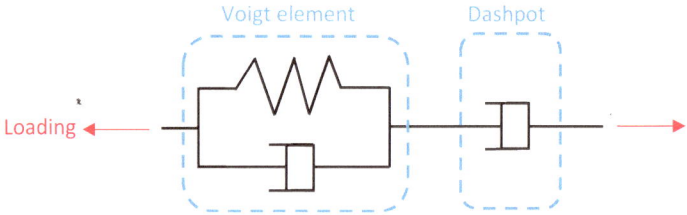

The deformation profile for this connection is shown below.

Strain γ

Constant
stress

Unloading

Non-linear relaxation
contributed by Voigt element

Non-linear creep
contributed by Voigt
element

Irreversible viscous strain
contributed by dashpot

Time

In summary, here are some key facts about viscoelastic fluids:

Typical physical states (Examples)	Liquid (polymeric systems)
Constitutive equation	Combination of both Newton's law (linear viscous part) and Hooke's law (linear elastic part), whereby the specific mix of both elements depends on the particular fluid. Examples of simple models used to fit viscoelastic data include the Maxwell and Voigt models Shear stress is a function of shear rate (or strain rate) from the viscous part and a function of strain from the elastic part $\tau = f(\dot{\gamma}, \gamma)$
Shear stress τ	Typically non-linear due to interplay between viscous and elastic properties May or may not be time-dependent, depends on whether the viscous part of the fluid mimics a time-dependent or time-independent fluid For example, in the case of a shear-thinning fluid, it would inherit time dependency if it mimics a thixotropic fluid for its viscous properties (as opposed to pseudoplastic)
Viscosity	Viscous (real fluid)
Elasticity	Has partial elastic properties, meaning some energy is stored, and some deformation is reversible
Viscosity function, η	Usually shear-thinning (can also be shear-thickening), depending on the type of fluid (e.g., either pseudoplastic or dilatant) its viscous part takes after

(i) Hookean Solid

Hookean solids are perfectly elastic solids much like an elastic metal spring. Their shear stress can be described using Hooke's law according to the equation $\tau = g\gamma$ whereby the shear stress scales linearly with strain (analogous to extension or displacement of a spring when stretched). g is the proportionality constant and represents the elastic constant.

Hookean spring

Under a compressive external force, the spring is pressed together and energy is stored in the system. When the external force is removed, the stored energy from the spring is used to return it back to its original (larger) length. The reverse happens with an extensional force that pulls the spring apart.

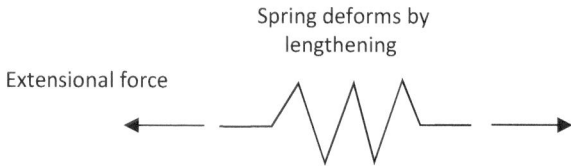

Key Features of the Hookean Spring

For an inviscid, perfectly elastic solid under an applied external force (e.g., shear stress), all the energy applied to the system is stored within the system as the material is deformed. In other words, no energy is lost or dissipated as heat like viscous inelastic liquids. When the external force is removed, the elastic solid restores to its original shape fully using the energy stored during deformation; hence the deformation is completely reversed upon unloading.

In addition to a full restoration of original state, the perfectly elastic solid also responds to the applied force instantaneously which means that a step change application of force would instantly produce the corresponding deformation (no gradual increase), and this value will be maintained for as long as this applied force is held constant. Any subsequent changes to the magnitude of the force will produce an instantaneous mirroring effect on the deformation (or strain).

In summary, here are some key facts about perfectly elastic solids:

Typical physical states (Examples)	Solids (metal spring, certain rubbers)
Constitutive equation	Follows Hooke's law $\tau = g\gamma$, where g is a constant
Shear stress τ	Scales linearly with strain
Viscosity	Inviscid
Elasticity	Perfectly elastic, all energy stored, and none dissipated; hence solid returns fully to original shape after applied stress is released

Problem 2

Discuss real-life examples of fluid properties and their practical significance. Include in your discussion Bingham fluids, shear-thinning, and shear-thickening fluids.

Solution 2

Worked Solution

Fluid properties may be engineered to facilitate processing needs or improve user experience for consumer products. We may also observe some of these properties in nature or in our daily encounters.

Bingham Fluids (Characteristic Yield Stress)

Certain common consumer items such as facial creams and toothpaste have a thick paste-like consistency and do not flow like a typical fluid when at rest. These are examples of Bingham fluids which start to flow like a fluid only when an applied force exceeds the fluid's inherent threshold value, also known as its yield stress.

There are practical advantages with this property such as in the case of toothpastes, as it does not flow (and hence leak out) from the tube when held upside down, hence making it easier to store and transport. This happens because the downward gravitational force is not large enough to overcome the fluid's yield stress; hence it behaves like a solid and does not flow. However, the toothpaste flows like a fluid when one gently squeezes on the tube. In this instance, the applied force now exceeds the fluid's yield stress, allowing it to become fluid-like and flow, thus facilitating its release from the tube.

Shear-Thinning Fluids

Shear-thinning fluids are given this name as their viscosity decreases as shear rate or shear force increases. This phenomenon occurs at higher values of shear rate/force. One useful application of this property is in bottled ketchup, whereby the fluid flows more easily as the bottle is shaken hard, resulting in lowered fluid viscosity, hence allowing one to control the amount of ketchup to be added to food according to one's preference.

Another food product that exhibits shear-thinning is the familiar spread on toast, margarine. It is essentially an oil derivative comprising chains of hydrogenated fatty acids. Under zero to low shear, the fluid's viscosity is high, and hence it takes on a thick and firm consistency which facilitates neat storage and transport in containers. However, during the filling and packaging process of margarine into these containers, the fluid is subject to higher shear forces (or shear rate) such that its viscosity reduces and the fluid flows more easily into the containers immediately after production.

Another application of shear-thinning occurs in paint, which is typically a suspension of solid pigments in a solvent. At low shear, such as when subject to gravitational force alone, paint maintains its high viscosity and displays a thick solid-like consistency. This makes it easy for storage and causes less mess. However, during painting, the paint fluid is sheared as it is brushed against a surface. This action reduces its viscosity, making it more fluid-like and able to mix effectively into a uniform mixture. The paint mixture can then be effectively spread over the surface. When brushing stops, viscosity returns to a higher value as shear reduces to zero. This means the paint becomes firmer and is less likely to sag or drip downwards due to gravity (e.g., when painting vertical walls). The relatively slow recovery of viscosity (due to thixotropic effect) of paint also provides sufficient time for the paint to flow and level before "firming up," hence minimizing the appearance of brush marks.

Fluids may also display shear-thinning if the duration of applied shear is prolonged (thixotropic). For thixotropic fluids that are time-dependent in shear-thinning, the longer the duration of application of applied shear, the lower their viscosity. These fluids become more liquid-like if a *sustained* force is applied. Some common examples include ketchup and nail polish which get "thinner" in consistency as you shake the bottle for a longer period of time. They return to their higher viscosity after the applied stress is removed.

Another example is honey, which becomes very viscous and "thick" especially at the bottom of the jar when we do not use it for a long period of time. This can be easily overcome by stirring honey, which effectively applies a higher shear force/rate, hence turning it into a less viscous and more easily flowing fluid.

Shear-Thickening Fluids

Shear-thickening fluids are the opposites of pseudoplastics in that they display higher viscosity when the magnitude of applied stress is high. One real-life example is wet sand, which stiffens into a "harder" or more viscous substance when applied force is large, such as when we run fast on a surface of wet sand. However, if we tread slowly and gently on wet sand, we find our feet start to sink down into it more easily as it behaves more fluid-like under a small applied force or stress.

We observe shear-thickening in quicksand as well, which is typically found in marshes near lakes or coastal areas. In terms of chemical structure, quicksand is essentially a colloidal gel that comprises of granular sand or silt suspended in water. The shear-thickening property is apparent when one is trapped in quicksand and tries to struggle out of it, which worsens the situation as the sand becomes "firmer." This is because the viscosity of quicksand increases under rapid or panicked movement when there is an increase in shear rate and applied force.

Another useful application of shear-thickening fluids is as an additive to body armors and bulletproof vests for self-defense purposes. Due to the increased viscosity under high shear stress or force, when an object such as a speeding bullet hits the armor at high force, the fluid shear thickens and becomes more rigid and solid-like, holding the fibers of the amour material together, hence increasing the overall strength of the material. This makes it harder for the bullet to penetrate far into the armor, and this offers protection to the body behind the armor. When the armor is not being penetrated by bullets, the wearer is able to move easily since the fluid's viscosity returns to its original lower value, making the armor material "less hard" and more flexible. However, one downside is that the armor would be less effective against slow piercing or slowly penetrating objects since the relatively low impact force does not cause much increase in viscosity and resultant stiffening of the material.

Other day-to-day examples of shear-thickening fluids include whipped cream and egg white. These fluids are time-dependent and shear-thickening (rheopectic) such that viscosity not only increases with shear rate but also with the duration of application of applied stress/force. Rheopectic fluids become denser as they are stressed over a period of time, but the force has to be sustained. A good example is egg white. When we whip egg whites, we can achieve a firmer (more viscous) cream-like consistency as we whip it over a longer period. The longer we whip, the stiffer the end product.

Problem 3

Discuss the molecular changes that occur in a shear-thinning and shear-thickening fluid under different flow conditions.

Solution 3

Worked Solution

The viscosity of real fluids often varies with shear rate, giving rise to interesting fluid properties. This phenomenon can be better understood when one studies the underlying molecular changes that occur to fluid particles under shear.

Shear-Thinning Fluids

Shear-thinning fluids exhibit a decreasing viscosity with increasing shear rate (let us call this the "ascending path" here for easy reference). When shear rate is reduced back to its original higher value, viscosity may return along the same path and at the same rate as the ascending path, back to its original higher value (i.e., pseudoplastic shear-thinning), or return to its original value via a different path and at a slower rate (i.e., thixotropic shear-thinning).

Pseudoplastic fluids are usually polymeric systems that consist of many coiled chains at the molecular level. When the polymers are at rest under no external shear, intermolecular forces present between the chains stabilize the fluid's structure, forming a structured fluid. However, under an applied shear force that pulls apart and stretches the chains, they begin to straighten and untangle, consequently aligning along the axis of shear and causing the internal fluid friction to decrease due to this alignment. This results in the observed reduced viscosity as shear rate increases.

Shear-thinning fluids may also display time-dependent viscosity, and this is commonly observed for fluids with microstructures such as suspended flocs, chain entanglements, and other loose molecular associations. Such fluids exhibit time dependency as a result of a competing effect between the "breakup" of microstructures by shearing stresses and their subsequent "rebuild" when shearing is reduced or stopped. Under zero shear, Brownian motion and random diffusion are at play, whereby resultant particle collisions help restore the fluid's microstructure to its original arrangement, and this takes place over a finite time. It follows that larger particles take longer to rebuild since diffusion rate is inversely proportional to particle size. Note that this time-dependent nature is more observable if the experimental time scale is significantly longer than the response time of the measuring instrument.

One example of a shear-thinning fluid with biological importance is triglyceride, a compound found in fats derived from animal or plant oil sources. In terms of chemical structure, triglyceride is an ester chemically produced via the reaction of three hydroxyl groups (-OH functional group) in the glycerol molecule and fatty acid groups (-COOH functional group) in the long-chain hydrocarbon molecules. The shear-thinning property of this fluid is largely contributed by hydrogen bonds that form between the fatty acid chains of neighboring molecules under zero or low shear. The triglyceride molecules are loosely held together via these bonds, forming a three-dimensional "network" structure that extends throughout the continuous phase of the system. This structure thereby gives the fluid substantial viscosity. The loose structure also contains void spaces that may nest even more molecules within them (e.g., additives like pigment molecules).

In general, hydrogen bonds are weak intermolecular forces that can be easily broken under shearing. When the fluid is subject to shear forces, these bonds are broken and the "network" collapses. The triglyceride chains are better able to slide past each other, and the fluid flows more easily, resulting in an overall reduced viscosity. These hydrogen bonds may be reformed when shear force (or shear rate) reduces back to a low value or zero, in turn reversing the shear-thinning effect, and viscosity increases back to its original higher value under zero or low shear.

Shear-Thickening Fluids

Shear-thickening fluids are the opposite, showing an increased viscosity with increased shear rate. Such fluids are also called dilatants since their volumes increase with increased shearing. At the molecular level, such fluids consist of a large percentage of dispersed and deflocculated particles such that there is just enough of the solvent phase (or liquid phase) for the particles to slide over one another at low shear.

When shear rate increases, the particles become more spaced out, giving rise to more air pockets and an increase in void volume, hence interparticle spacing. This makes the system take on the consistency of a thicker and firmer viscous paste under high shear rates.

Problem 4

Fluids may display Newtonian or non-Newtonian behavior. For Newtonian fluids such as water, there is a linear dependence between shear stress, τ, and shear rate, $\dot{\gamma}$. This linear dependence is commonly represented by η a constant viscosity value in the mathematical expression $\tau = \eta\dot{\gamma}$.

However, for many commercial polymer melts, colloidal suspensions, and other viscous suspensions, they display non-Newtonian behavior. Explain what is meant by non-Newtonian behavior, and include appropriate graphs showing the relationship between shear stress and shear rate, as well as that between viscosity and shear rate for such fluids.

Solution 4

Worked Solution

For a Newtonian fluid, viscosity is constant even if shear rate changes. Its constitutive equation can be expressed in the form as shown below.

$$\tau = \eta\dot{\gamma}$$

$$\tau(t) = \eta\dot\gamma(t), \eta \text{ is constant}$$

In the equation above, we notice two main features. First, the stress response τ scales linearly with the applied shear rate $\dot\gamma$ since the scaling factor η is a constant. Secondly, the stress response is only dependent on the instantaneous shear rate, i.e., the value of τ at a specific time t, and therefore it is a direct mapping of the applied shear rate at that instant of time t. In other words, the shear stress response is a mathematical function with variable t, and this function follows the time-dependent function that describes the applied shear rate $\dot\gamma(t)$.

In a non-Newtonian fluid, the local shear stress τ and local shear rate $\dot\gamma$ have a non-linear relationship, which means the constitutive equation used for Newtonian fluids ($\tau = \eta\dot\gamma$ whereby η is a constant value) cannot be used to model these fluids. For non-Newtonian fluids, viscosity η is not a fixed scalar quantity but a variable, and this variable may depend on shear rate *as well as* the time history of shear rate.

If we were to use the same equation $\tau = \eta\dot\gamma$ to model a non-Newtonian fluid, we can modify it such that η is replaced by an apparent (or observed) viscosity, η_{app}. The constitutive equation becomes $\tau = \eta_{app}\dot\gamma$, whereby η_{app} is no longer a constant but a function of other variables. Some examples of non-Newtonian fluids include ketchup, toothpaste, custard, shampoo, paint, and even human blood.

Non-Newtonian behavior can exist in the following forms:

- *Shear-thinning*

  Viscosity decreases with increasing shear rate: $\eta_{app} \to \eta_{app}(\dot\gamma)$. There is an additional dependence of shear stress on shear rate ($\tau(t) \to \tau(t,\dot\gamma)$) which originates from the dependence of viscosity on shear rate. In totality, we see that the constitutive equation takes on the following form:

$$\tau = \eta_{app}\dot\gamma$$

$$\tau(t,\dot\gamma) = \eta_{app}(\dot\gamma)\dot\gamma(t)$$

  Examples of such fluids include nail polish and whipped cream.

- *Shear-thickening (also called "dilatants")*

  Viscosity increases with increasing shear rate: $\eta_{app} \to \eta_{app}(\dot\gamma)$. Like shear-thinning fluids, there is an additional dependence of shear stress on shear rate ($\tau(t) \to \tau(t,\dot\gamma)$). It follows that the constitutive equation becomes:

$$\tau = \eta_{app}\dot\gamma$$

$$\tau(t,\dot\gamma) = \eta_{app}(\dot\gamma)\dot\gamma(t)$$

  Even though the equation above looks the same as that for shear-thinning fluids, there is a difference in the exact mathematical expression for the function of apparent viscosity, $\eta_{app}(\dot\gamma)$. Examples of such fluids include thickening agents and corn starch solution.

- *Bingham fluid*

     For this type of fluid, there is a minimum stress called the "yield stress" denoted τ_y that must be exceeded for the fluid to flow or deform. When the applied stress is less than the yield stress, the fluid material behaves like a rigid body and is highly elastic (or "stiff"). When applied stress is more than yield stress, the fluid material begins to flow, and in this state, the relationship between stress and shear rate may be either linear (Newtonian-like with constant viscosity) or non-linear (non-Newtonian shear-thinning or shear-thickening).

The properties of the abovementioned types of non-Newtonian fluids are shown in the two graphs below, with reference to the behavior of a Newtonian fluid.

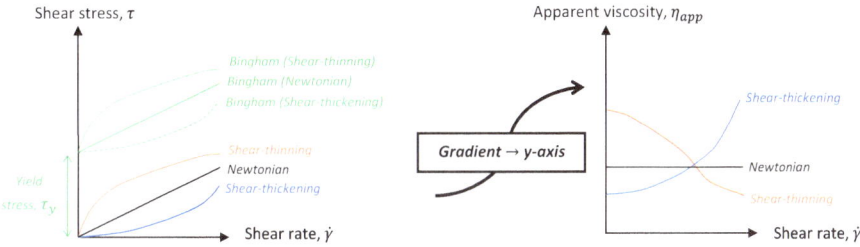

Flow Models and Constitutive Equations

Problem 5

Explain what is meant by a constitutive equation, and provide examples of some typical models used to represent it.

Solution 5

Worked Solution

A constitutive equation is of importance in rheological studies as it relates shear stress τ to shear rate $\dot{\gamma}$. This equation can be used to describe flow behavior such as how viscosity changes under different flow conditions (e.g., varying shear rates). Different flow models have since been developed to describe a wide range of fluids, from the simpler ones that are more suitable for narrow ranges of shear rates (e.g., Newtonian and power law models) to the more complex ones (e.g., multiparameter models).

Simple models are called as such because they typically show simple linear behavior, which in graphical terms means that the model equations produce straight-line plots on linear or logarithmically scaled graphs. Other more rigorous models consist of more parameters and variables. Although this poses greater mathematical challenge, these models can fit a wider range of data points covering diverse flow conditions. This is often the case for real fluids or real-life operating conditions. One example is the Cross model which comprises four parameters.

The table below gives a brief snapshot of common types of constitutive equations that model different types of fluid behavior. Note that while flow models may be categorized discretely as shown below, real fluids are usually best characterized by combining different models. It is worth noting that one should exercise some

X. W. Ng, *Pocket Guide to Rheology: A Concise Overview and Test Prep for Engineering Students*, https://doi.org/10.1007/978-3-030-30585-7_2

flexibility in applying elements of different models to arrive at the best-fit constitutive equation to describe a fluid (e.g., we may combine the Bingham model with the Cross model, etc.).

Flow model	Constitutive equation	No. of parameters	Notes
Newtonian	$\tau = \eta\dot{\gamma}$	1 (η)	Simplest of all models Constant viscosity η, independent of shear rate (hence termed "Newtonian") Single data point sufficient
Power law	$\tau = k\dot{\gamma}^n$	2 (k, n)	$n = 1$ returns the Newtonian case Non-Newtonian if $n \neq 1$, viscosity is not constant Viscosity decreases with shear rate for shear-thinning ($n < 1$) and increases for shear-thickening ($n > 1$) Curve fitting requires more than one data point (more than one parameter in the model) Straight-line plot on logarithmic scale
Bingham plastic	$\tau = \tau_y + \eta\dot{\gamma}$	2 (τ_y, η)	Bingham means a characteristic yield stress τ_y exists, below which which no flow occurs (solid-like) Above τ_y, a fluid can be Newtonian or non-Newtonian (shear-thinning or thickening). "Bingham plastic" here refers to the case of a Newtonian Bingham fluid Sum of Newtonian model and a yield stress term
Herschel Bulkley	$\tau = \tau_y + k\dot{\gamma}^n$	3 (τ_y, k, n)	Bingham-type fluid with a characteristic yield stress, below which no flow occurs (solid-like) Above τ_y, this model can predict both Newtonian when $n = 1$ (back to case of Bingham plastic model) and non-Newtonian when $n \neq 1$ For $n < 1$, shear-thinning case which is also known as "Bingham viscoplastic" (viscous, inelastic, shear-thinning) For $n > 1$, shear-thickening case Sum of power law model and a yield stress term
Cross	$\tau = \left[\eta_\infty + \dfrac{\eta_0 - \eta_\infty}{1+(\alpha\dot{\gamma})^m}\right]\dot{\gamma}$	4 (α, η_0, η_∞, m)	Often used for modelling real fluids as it can be applied for low and high shear rates η_0 and η_∞ denote asymptotic values of viscosity at very low and very high shear rates, respectively. The other parameters α and m are constants

Problem 6

Given that a particular Maxwell model comprises of a single dashpot and a single spring connected in series as shown in the illustration below, show that the strain ε experienced by a fluid modelled using this Maxwell model under a constant applied shear stress σ may be expressed as follows:

$$\varepsilon = \sigma\left(\frac{1}{g} + \frac{t}{\eta}\right)$$

Assuming that the spring element obeys Hooke's law and has an elastic constant denoted g and the dashpot element is an ideal Newtonian fluid with a constant viscosity η, derive the stress profile for the Maxwell fluid as a function of time. Identify any key parameters or quantities in the expression, and briefly comment on their significance if any.

Solution 6

Worked Solution

Let us first examine the strain experienced by the spring component of the Maxwell model.

Under a constant applied stress, the elastic response of the spring can be described using Hooke's law as follows:

$$\sigma_s = \varepsilon_s g \rightarrow \varepsilon_s = \frac{\sigma_s}{g}$$

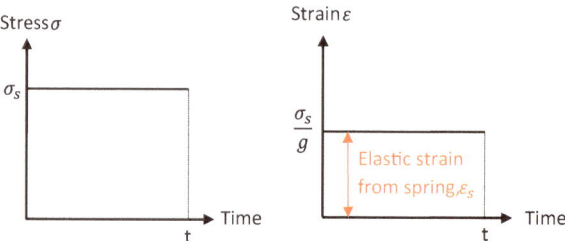

Let us now examine the strain experienced by the dashpot component.

Under a constant applied stress, the viscous response of the dashpot can be described using Newton's law as follows:

$$\sigma_d = \eta \dot{\varepsilon}_d = \eta \frac{d\varepsilon_d}{dt} \rightarrow \frac{d\varepsilon_d}{dt} = \frac{\sigma_d}{\eta}$$

$$\varepsilon_d = \frac{\sigma_d}{\eta} t$$

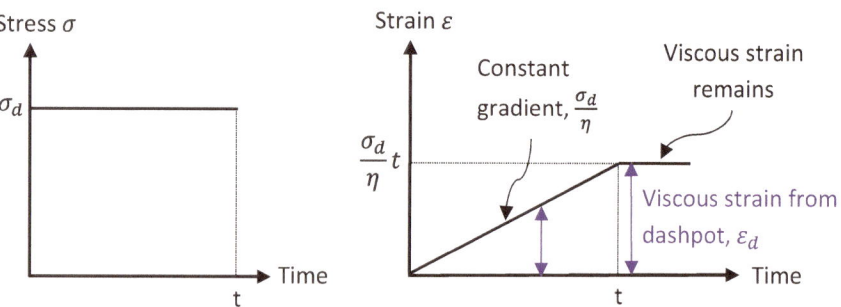

Now when we add both elements together in series, we can derive the overall strain response as follows:

$$\varepsilon = \text{elastic strain} + \text{viscous strain}$$

$$\varepsilon = \varepsilon_s + \varepsilon_d = \frac{\sigma_s}{g} + \frac{\sigma_d}{\eta} t$$

We know that in a series connection, along the axis which the stress acts, the applied stress is constant. Therefore, the total stress in a Maxwell element, denoted here as σ, is equivalent to the individual stresses experienced by the spring element and dashpot element, respectively.

$$\sigma = \sigma_s = \sigma_d$$

We can now derive the given expression for overall strain experienced by the Maxwell element.

$$\varepsilon = \sigma\left(\frac{1}{g} + \frac{t}{\eta}\right)$$

We can also illustrate the strain and stress profiles for the Maxwell element in the plot as shown below.

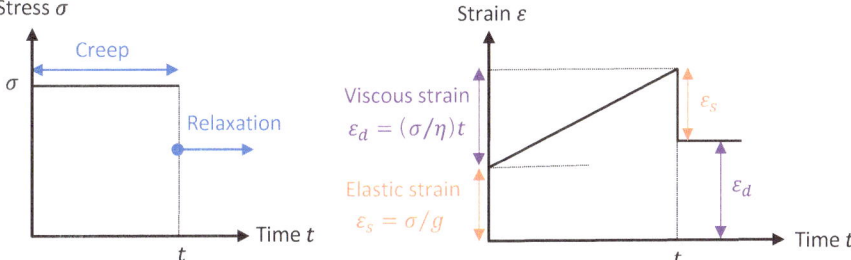

There are two key phases to note:

1. *Creep* – Stress is applied.
2. *Relaxation* – Stress is removed (after a prior period of stress application).

Strain Profile

During creep, the total strain is a sum of elastic strain and viscous strain. Note that during relaxation, total strain is equivalent to viscous strain only, as elastic strain is zero.

$$\varepsilon\big|_{\text{creep}} = \varepsilon_s + \varepsilon_d$$

$$\varepsilon\big|_{\text{relaxation}} = \varepsilon_d$$

Let us derive an expression for the stress relaxation phase. For a series connection, we note that $\sigma_s = \sigma_d$; therefore we have the following starting differential equation.

$$g\varepsilon_s = \eta \frac{d\varepsilon_d}{dt}$$

We know that the strains contributed by the dashpot and spring elements sum up to give the overall total strain of the Maxwell element; therefore we express total strain ε as follows:

$$\varepsilon = \varepsilon_d + \varepsilon_s \rightarrow \varepsilon_d = \varepsilon - \varepsilon_s$$

$$\frac{d\varepsilon_d}{dt} = \frac{d(\varepsilon - \varepsilon_s)}{dt} = \frac{d\varepsilon}{dt} - \frac{d\varepsilon_s}{dt}$$

$$g\varepsilon_s = \eta \left(\frac{d\varepsilon}{dt} - \frac{d\varepsilon_s}{dt} \right)$$

Stress Profile

In order to derive the stress relaxation profile, we may first notice that during relaxation, since applied stress is zero, the total strain of the system is constant with respect to time, i.e., $\frac{d\varepsilon}{dt}$ is zero. This is also evident from the plots above for the relaxation period. Therefore, we may impose this condition and simplify the earlier expression to the following:

$$g\varepsilon_s = -\eta \frac{d\varepsilon_s}{dt}$$

$$\frac{d\varepsilon_s}{\varepsilon_s} = -\frac{g}{\eta} dt$$

$\ln \varepsilon_s = -\frac{g}{\eta} t + \ln A$, where A is an integration constant

$$\varepsilon_s = A e^{-\frac{g}{\eta} t}$$

At this point we recall that for a series connection, $\sigma = \sigma_s = \sigma_d$. For ease of mathematical simplification, we use the relationship $\sigma = \sigma_s = g\varepsilon_s$ as shown below.

$$\sigma = g\varepsilon_s = gA e^{-\frac{g}{\eta} t}$$

When $t = 0$, we obtain the following by substituting into the earlier expression as follows:

$$\varepsilon = \sigma\left(\frac{1}{g} + \frac{0}{\eta}\right) = \frac{\sigma}{g}$$

$$\sigma = g\varepsilon \rightarrow A = \varepsilon$$

Therefore, we can express stress for an arbitrary time t as follows:

$$\sigma(t) = g\varepsilon e^{-\frac{g}{\eta}t} = g\varepsilon e^{-\frac{t}{\lambda}}, \quad \text{where } \lambda = \eta/g$$

The significance of the quantity λ is that it represents the "relaxation time" of the viscoelastic fluid. It is also equivalent to the time taken for the value of stress σ to reduce to $1/e$ times its initial value at the beginning of the relaxation phase, which corresponds to when $t = \eta/g$.

$$\sigma(t = \eta/g) = g\varepsilon e^{-1}$$

$$\sigma(t = 0) = g\varepsilon$$

$$\therefore \sigma(t = \lambda) = \frac{\sigma(t = 0)}{e}$$

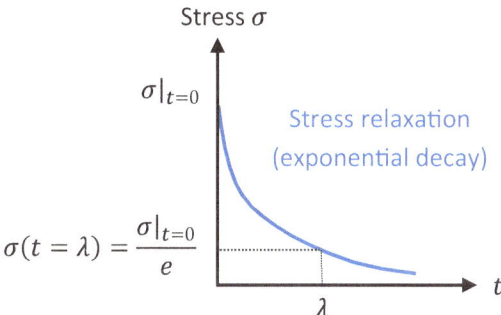

Problem 7

The Burger's model is sometimes used to characterize emulsion systems. It consists of a linear combination of a Kevin-Voigt element and a Maxwell element as shown below.

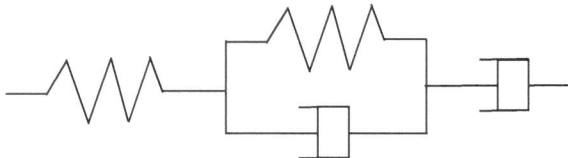

(a) **Discuss any advantages of this model as compared to a single Maxwell or a single Voigt element, and comment on how it may be improved to apply to a wider range of fluids.**

(b) **Plot the expected shape of the strain versus time graph for the Burger's model, and comment on key features of the plot in relation to the viscous and elastic components of the model.**

(c) **The rheological properties of certain viscoelastic food products such as ice cream may be studied using this model. An understanding of how fluid properties (e.g., viscosity) change with varying parameters (e.g., concentration of additives such as stabilizing agents or thickeners) will help food engineers achieve desired product specifications such as a smooth texture and suitable melting point for ice cream.**

  (i) **Given that the Maxwell element consists of a spring component with elasticity g_1 and strain of γ_1 and a dashpot component with viscosity η_1 and strain of γ_3 and that the Voigt element consists of a spring component with elasticity g_2 and dashpot component with viscosity η_2 and strain of γ_2, derive the following differential equation for the Voigt element of the Burger's model.**

$$\eta_2 \frac{d\gamma_2}{dt} = -g_2\gamma_2 + \sigma$$

  (ii) **Given the solution for a general first-order differential equation as shown below**

$$A\frac{dy}{dx} = -y + B \rightarrow y = B\left(1 - e^{-\frac{x}{A}}\right)$$

  Derive the total strain of the Burger's model as follows where $\lambda = \frac{\eta_2}{g_2}$:

$$\gamma = \sigma\left[\frac{1}{g_1} + \frac{1}{g_2}\left(1 - e^{-\frac{t}{\lambda}}\right) + \frac{t}{\eta_1}\right]$$

  (iii) **Revisiting the plot in part (b), plot the graphs of the individual component strains contributed by the Maxwell spring, Maxwell dashpot, and Voigt element in arriving at the final total strain curve.**

Solution 7

Worked Solution

(a) The Burger's model is an improvement from a single Maxwell element and a single Voigt element as it combines the advantages of both elements in a single model. One of the key downsides of a Maxwell element lies in its inability to model non-linear strain creep due to the series connection between one dashpot and one spring. As for the Voigt element, it is able to model non-linear strain growth but is less ideal in modelling the partial irreversible deformation caused by the viscous part of a viscoelastic fluid due to the parallel connection which forces all deformation to be completed reversed (or restored) upon removal of applied stress.

The Burger's model allows us to visualize three key deformations:

1. Instantaneous elastic deformation from the Maxwell spring
2. Irreversible deformation from the Maxwell dashpot
3. Non-linear deformation changes with the application and removal of an external stress from the Voigt model

The Burger's model consists of a single Maxwell and a single Voigt element. The model can be varied and adjusted by including multiple numbers of each element (e.g., multiple elastic springs, each with a different elastic constant, or multiple dashpots each with a different viscosity) and/or by having different combinations of the constituent elements (e.g., two Maxwell elements, three Voigt elements, and one dashpot connected in series). The infinite number of ways that one can combine various elements together allows us to model a wide range of viscoelastic fluids.

One example of a plausible combination is by connecting multiple Maxwell elements (say for an arbitrary number of n elements) in parallel whereby the total stress profile will then be the sum of the individual stresses in each branch. For a single Maxwell element, we have the stress profile as follows.

$$\sigma(t) = g\gamma e^{-\frac{t}{\lambda}}, \quad \text{where } \lambda = \eta/g$$

Using an arbitrary example where we have three elements (i.e., $n = 3$), we have the following illustration and expression for total stress.

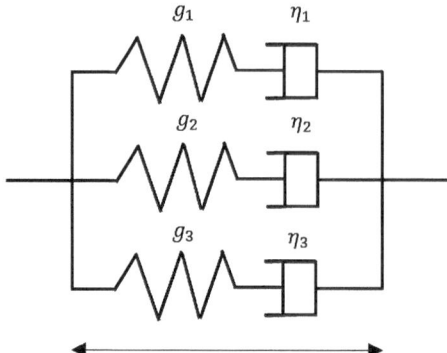

All Maxwell elements undergo same
strain γ due to parallel connection

$$\sigma(t) = g_1\gamma e^{-\frac{t}{\lambda_1}} + g_2\gamma e^{-\frac{t}{\lambda_2}} + +g_3\gamma e^{-\frac{t}{\lambda_3}}$$

We can proceed to define a generalized model for this type of combination, for n Maxwell elements where each element has its own (and may differ) relaxation time $\lambda_i = \eta_i/g_i$.

$$\sigma(t) = \gamma \sum_{i=1}^{n} g_i e^{-\frac{t}{\lambda_i}}$$

It is worth noting that the Burger's model has been shown to be best used for viscoelastic colloidal and emulsion-type systems with firm gel-like texture and shear-thinning properties. Colloidal systems may comprise of substances dissolved in solution and/or particles dispersed in a suspension.

(b) The following graph shows the strain (or deformation) against time over an initial period of constant stress σ (before the vertical dotted line), followed by removal of this stress (after the vertical dotted line).

Burger's Model

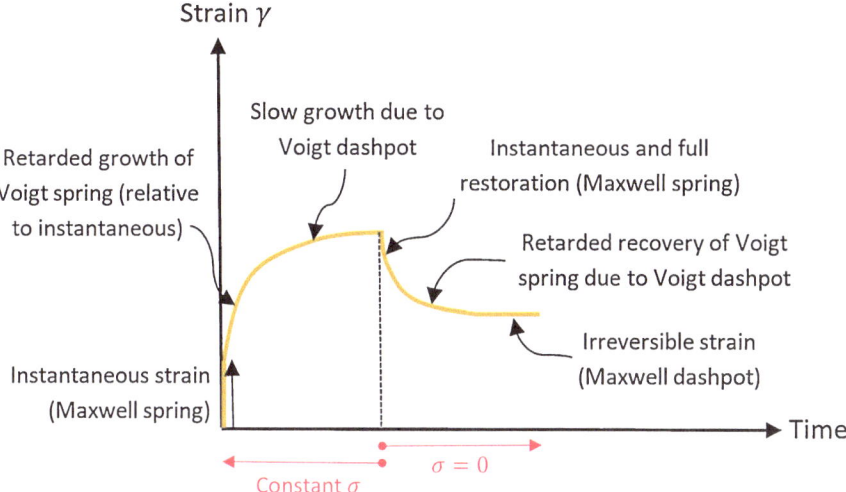

(c) (i) Below is an illustration of the Burger's model.

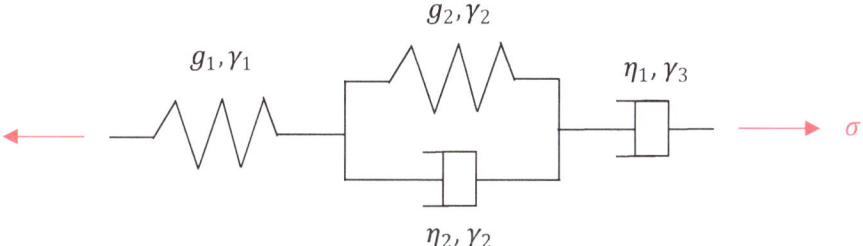

We can express the elastic deformation (instantaneous) of the Maxwell spring with elastic constant g_1 as follows.

$$\gamma_1 = \frac{\sigma}{g_1}$$

We can also express the viscous deformation of the Maxwell dashpot with viscosity η_1 as follows.

$$\gamma_3 = \frac{\sigma t}{\eta_1}$$

Note that in a series connection, the spring and dashpot act independently in their contributions to strain. This is unlike the Voigt element whereby the parallel connection of the dashpot and spring forces both to experience the same strain (stretch or deform the same amount).

Let us denote the deformation of the Voigt element as γ_2. For this element, stresses in each parallel branch sum up to give the overall stress σ.

$$\sigma = \sigma_{\text{Voigt dashpot}} + \sigma_{\text{Voigt spring}}$$

$$\sigma_{\text{Voigt spring}} = g_2 \gamma_2 \tag{1}$$

$$\sigma_{\text{Voigt dashpot}} = \eta_2 \frac{d\gamma_2}{dt} \tag{2}$$

Substituting the above expressions 1 and 2 into the stress equation, we have the following differential equation for the strain in the Voigt element:

$$\sigma = \eta_2 \frac{d\gamma_2}{dt} + g_2 \gamma_2$$

$$\eta_2 \frac{d\gamma_2}{dt} = -g_2 \gamma_2 + \sigma$$

$$\left(\frac{\eta_2}{g_2}\right) \frac{d\gamma_2}{dt} = -\gamma_2 + \frac{\sigma}{g_2}$$

$$\lambda \frac{d\gamma_2}{dt} = -\gamma_2 + \frac{\sigma}{g_2}$$

The above is a familiar first-order differential equation commonly encountered in engineering problems. The solution can be mathematically proven (not done in this problem but shown in subsequent problems – refer to integrating factor) to be as follows which is also given in the problem.

$$\gamma_2 = \frac{\sigma}{g_2}\left(1 - e^{-\frac{t}{\lambda}}\right)$$

where the characteristic timescale also known as relaxation time λ is defined as $\lambda = \frac{\eta_2}{g_2}$.

(c) (ii) Having derived the strain of the Voigt element in part (c) (i), we can now easily derive the expression for the total deformation γ experienced by the Burger's model by summing the individual component strains.

γ_1 comes from the Maxwell spring and gives rise to the initial instantaneous elastic strain; γ_2 comes from the spring and dashpot of the Voigt element and contributes to the exponential profile for deformation (reversible) during the strain growth and decay periods. And finally, γ_3 comes from the Maxwell dashpot and explains the irreversible deformation that can be observed at final steady state.

$$\gamma = \gamma_1 + \gamma_2 + \gamma_3$$

$$\gamma = \frac{\sigma}{g_1} + \frac{\sigma}{g_2}\left(1 - e^{-\frac{t}{\lambda}}\right) + \frac{\sigma t}{\eta_1}$$

$$\gamma = \sigma\left[\frac{1}{g_1} + \frac{1}{g_2}\left(1 - e^{-\frac{t}{\lambda}}\right) + \frac{t}{\eta_1}\right]$$

(c) (iii) Revisiting the plot in part (b), we can now plot the graphs of the individual strains contributed by the components of the Burger's model, namely, the Maxwell spring, the Maxwell dashpot, and the Voigt element. Visually, we can see how the final total strain curve was derived in part (b), as a sum of the three graphs for γ_1, γ_2, and γ_3 as shown below.

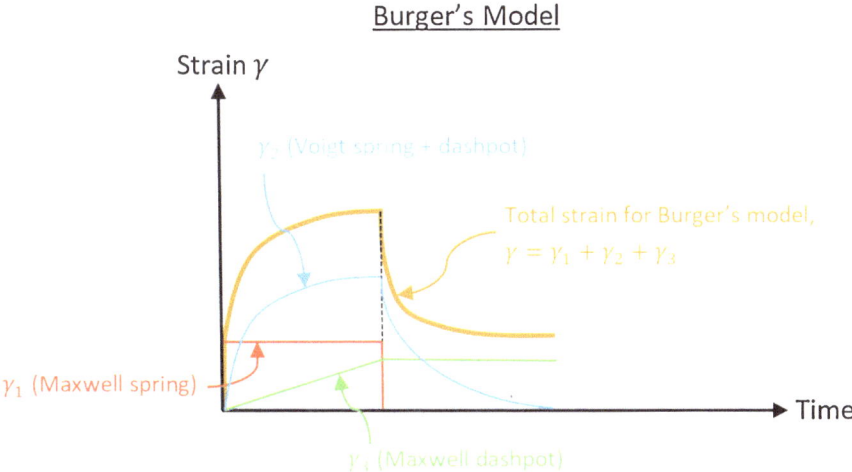

Problem 8

There are various models developed to predict viscoelastic behavior of fluids such as polymer melts. One such model is the Maxwell model which couples an elastic component with a viscous component and can be likened to a linear spring and a linear dashpot system connected in series as illustrated below.

(a) **Derive the constitutive equation for a Maxwell fluid as shown below, where**
     λ denotes relaxation time. Briefly explain what relaxation time means and
     how it varies for different fluids.

$$g\frac{d\gamma}{dt} = \frac{d\tau}{dt} + \frac{\tau}{\lambda}$$

(b) **Using results in part (a), derive the expression below for the response of**
     shear stress under stress relaxation (after steady shear). Plot graphs of
     strain rate and shear stress over time for this response.

$$\tau = \tau_0 e^{-\frac{t}{\lambda}}$$

(c) **Using an appropriate integrating factor, derive the following integral strain**
     rate equation, where t' represents past time variable. Show that the same
     expression in part (b) can be obtained using the equation below.

$$\tau(t) = \int_{t'=-\infty}^{t'=t} g\dot{\gamma}(t')e^{-\frac{(t-t')}{\lambda}}dt'$$

(d) **Derive the following integral strain equation for this problem using inte-**
     gration by parts on the strain rate equation in part (c).

$$\tau(t) = -\int_{-\infty}^{t'} \gamma\left[\frac{g}{\lambda}e^{-\frac{(t-t')}{\lambda}}\right]dt'$$

   Show that the same result in part (b) can be obtained using this integral
   strain rate equation. Plot a graph of strain against time.

Solution 8

Worked Solution

(a) Following from the spring and dashpot analogy for a Maxwell fluid, we can
    express shear stress for each component. Let τ_1 and τ_2 denote the shear stresses
    for the spring and dashpot components, respectively. The elasticity of the
    polymer chain is likened to that in a spring, while viscosity can be represented
    by the dashpot element.

For the spring component, we have elastic constant g and strain γ_1, while for the dashpot component, we have viscosity η and strain rate $\dot{\gamma}_2$.

$$\tau_1 = g\gamma_1$$
$$\tau_2 = \eta\dot{\gamma}_2$$

Due to the continuity of stress in a series system, we can establish that total stress τ is equivalent to stress in each component.

$$\tau = \tau_1 = \tau_2$$

$$\gamma_1 = \frac{\tau}{g} \tag{1}$$

$$\dot{\gamma}_2 = \frac{\tau}{\eta} \tag{2}$$

We can also infer from the connection in series that the total strain γ is a sum of the strains from each element (i.e., strain additivity):

$$\gamma = \gamma_1 + \gamma_2$$

It follows that strain rate would also be additive.

$$\dot{\gamma} = \dot{\gamma}_1 + \dot{\gamma}_2$$

The differential form of strain rate can be expressed as follows with a derivative with respect to time:

$$\dot{\gamma} = \frac{d\gamma}{dt} = \frac{d\gamma_1}{dt} + \dot{\gamma}_2$$

Substituting expressions 1 and 2, we obtain a first-order ordinary differential equation representing the constitutive equation for a Maxwell fluid. In this equation, we have defined a new quantity, relaxation time $\lambda = \eta/g$. Note that if the fluid was fully elastic, i.e., no viscous component, then $\lambda \to \infty$. Conversely, if the fluid was fully viscous, then $\lambda \to 0$.

$$\frac{d\gamma}{dt} = \frac{1}{g}\frac{d\tau}{dt} + \frac{\tau}{\eta}$$

$$g\frac{d\gamma}{dt} = \frac{d\tau}{dt} + \frac{\tau}{\lambda}$$

(b) During stress relaxation, $\dot{\gamma} = \frac{d\gamma}{dt} = 0$, therefore, our result in part (a) simplifies as follows:

$$g\frac{d\gamma}{dt} = \frac{d\tau}{dt} + \frac{\tau}{\lambda} = 0$$

$$\frac{d\tau}{dt} = -\frac{\tau}{\lambda}$$

In order to integrate the differential equation, we need to establish boundary conditions. One of the boundary conditions is the initial condition at $t = 0$ when we have instantaneous steady shear and strain rate is a constant value, $\dot{\gamma}_0$. After $t = 0$, at stress relaxation occurs.

$$\frac{d\tau}{dt} = 0$$

$$\tau = \eta\dot{\gamma}_0 = \tau_0$$

We can now integrate our earlier differential equation under stress relaxation.

$$\int_{\tau_0}^{\tau} \frac{d\tau}{\tau} = -\int_0^t \frac{dt}{\lambda}$$

$$\ln\left(\frac{\tau}{\tau_0}\right) = -\frac{t}{\lambda}$$

$$\tau = \tau_0 e^{-\frac{t}{\lambda}}$$

We can visualize the flow condition by plotting the behaviors of shear stress and strain rate over time. We observe that there is an exponential decay of shear stress after strain rate turns zero at $t = 0$. Shear stress does not become zero instantly due to viscoelasticity of the fluid. The viscoelastic characteristic contributes to the non-Newtonian response.

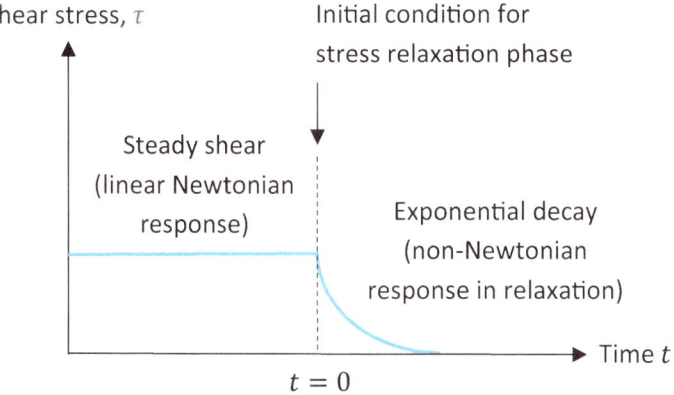

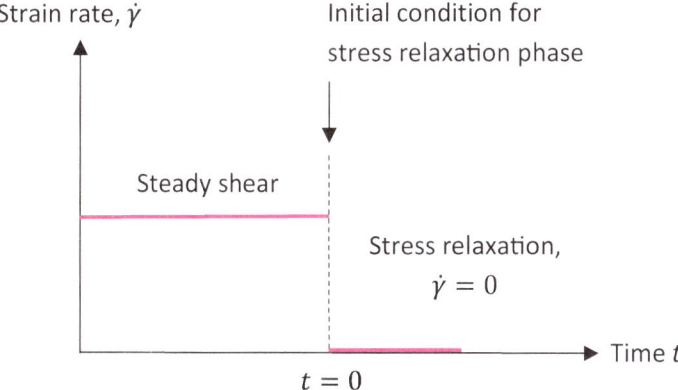

(c) Let us begin with our differential equation found earlier and consider a suitable integrating factor.

$$g\frac{d\gamma}{dt} = \frac{d\tau}{dt} + \frac{\tau}{\lambda}$$

$$\text{Integrating Factor} = e^{\int \frac{1}{\lambda}dt} = e^{\frac{t}{\lambda}}$$

Multiplying the differential equation throughout with the integrating factor (indicated in purple below), we have the following:

$$g\frac{d\gamma}{dt}(e^{\frac{t}{\lambda}}) = \frac{d\tau}{dt}(e^{\frac{t}{\lambda}}) + \frac{\tau}{\lambda}(e^{\frac{t}{\lambda}}) = \frac{d}{dt}(\tau e^{\frac{t}{\lambda}})$$

$$g\dot{\gamma}e^{\frac{t}{\lambda}} = \frac{d}{dt}\left(\tau e^{\frac{t}{\lambda}}\right)$$

$$\tau e^{\frac{t}{\lambda}} = \int g\dot{\gamma}e^{\frac{t}{\lambda}}dt$$

At this point we introduce a new variable for time called the *past time variable* denoted t' in order to perform the integration. This is necessary because shear stress at a current time t, denoted by $\tau = \tau(t)$ in the above expression, is affected by strain rate from the past (i.e., a time in history when current time t is undefined since it does not make sense to have negative values for t, so historical time can be represented by another variable t' instead). Past time variable t' therefore encompasses a wider time frame than current time t since t only starts from zero when the clock starts. On the other hand, t' is a "fictional" variable we have created, and it is able to take both positive and negative values (up to $-\infty$), and therefore it has a range of $-\infty < t' < 0$ which can effectively represent historical time as well.

Following from the above discussion, we will now replace our variable of t with t' for the integral term as shown below. Note that strain rate is not a constant but is a function of past time, i.e., $\dot{\gamma} = \dot{\gamma}(t')$. $\tau(t)$ on the left-hand side of the equation still refers to shear stress at current time t.

$$\tau e^{\frac{t}{\lambda}} = \int g\dot{\gamma}(t')e^{\frac{t'}{\lambda}}dt'$$

The limits of our integral based on the past time variable can go from history (lower limit of $-\infty$) to current time t.

$$\tau e^{\frac{t}{\lambda}} = \int_{t'=-\infty}^{t'=t} g\dot{\gamma}(t')e^{\frac{t'}{\lambda}}dt'$$

$$\tau(t) = \int_{t'=-\infty}^{t'=t} g\dot{\gamma}(t')e^{-\frac{(t-t')}{\lambda}}dt'$$

We have shown the equation given in part (c). Next, we need to use this equation to derive the same result as in part (b). To do so, let us consider defining a new variable for ease of mathematical treatment. Let $s = t - t'$ such that when $t' = t$, $s = 0$ and when $t' = -\infty$, $s = \infty$. Also, $ds = -dt'$.

With this change of variables, we now reduce the number of time variables from two (i.e., t, t') to just one (i.e., s). This simplifies our integral as shown below.

$$\tau = -\int_{s=\infty}^{s=0} g\dot{\gamma}(s)e^{-\frac{s}{\lambda}}ds$$

In our problem we have two distinct periods to consider:

1. *Steady shear* – Period of constant strain rate from historical time up to current time $t = 0$.
2. *Stress relaxation* – After steady shear, this is a period of relaxation when there is zero strain.

Variables	Steady shear	Stress relaxation
Strain rate, $\dot{\gamma}$	$\dot{\gamma}_0$	0
Past time, t'	$-\infty$ to 0	0 to t
New time variable, s	∞ to t	t to 0

$$\tau = \int_{\infty}^{t} -g\dot{\gamma}(s)e^{-\frac{s}{\lambda}}ds + \int_{t}^{0} -g\dot{\gamma}(s)e^{-\frac{s}{\lambda}}ds$$

For the first integral term during steady shear, we have a constant strain rate denoted $\dot{\gamma}_0$ below. Since it is a constant value, it can be taken out of the integral. As for the second integral term representing stress relaxation, the integral disappears from the equation since its value is zero, arising from a strain rate of zero during this period.

$$\tau = -\dot{\gamma}_0 \int_{\infty}^{t} ge^{-\frac{s}{\lambda}}ds + 0$$

$$\tau = g\lambda\dot{\gamma}_0\left[e^{-\frac{s}{\lambda}}\right]_{\infty}^{t} = \eta\dot{\gamma}_0 e^{-\frac{t}{\lambda}} = \tau_0 e^{-\frac{t}{\lambda}}$$

We have therefore derived the same result as in part (b) using the integral strain rate equation. We can plot the following graph of shear stress τ (blue line) against past time variable t'. The plot is also annotated to differentiate between the different time variables.

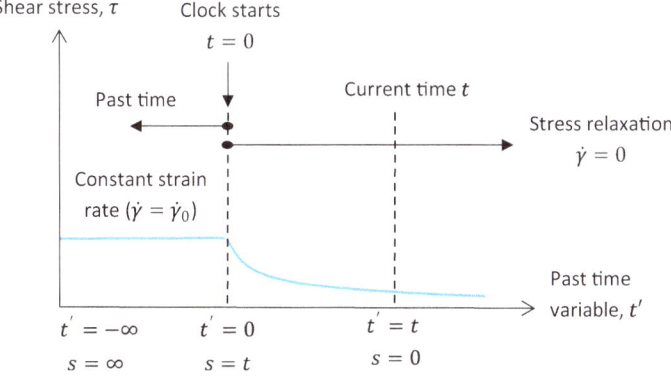

(d) Before we derive the integral strain equation, we should first observe that strain γ increases in value as we move backward in time. At current time t, strain is zero.

We note further that strain rate is zero from current time all the way backward in time until $t = 0$. This is the period of stress relaxation. Strain rate is the gradient of a plot of strain against time; hence the gradient for the stress relaxation period is zero, giving us a straight horizontal line. This horizontal line has a value of zero since strain is zero at current time t.

As we move further backward in time from $t' = 0$ to $t' = -\infty$, we have the period of constant strain rate $\dot{\gamma}_0$. This means that the gradient of this part of the plot of strain against time is a constant non-zero value equivalent to $\dot{\gamma}_0$, giving rise to a straight line with positive gradient (assuming $\dot{\gamma}_0 > 0$).

Finally, the two lines should connect at the point $t' = 0$ for continuity; hence we arrive at the graph shown below.

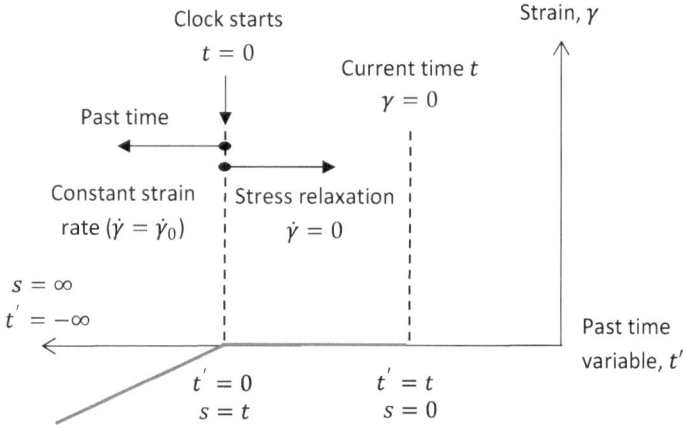

To derive the integral strain equation, let us consider the integral strain rate equation from part (c).

$$\tau(t) = \int_{-\infty}^{t} g\dot{\gamma}(t')e^{-\frac{(t-t')}{\lambda}}dt' = \int_{-\infty}^{t}\left[ge^{-\frac{(t-t')}{\lambda}}\right]\left[\frac{d\gamma}{dt'}\right]dt'$$

We can use integration by parts on the expression above. Recall that integration by parts is defined as follows where u and v are two functions of the same variable.

$$\int u\,dv = uv - \int v\,du$$

In our expression, we observe two such functions of t' represented by square brackets. Hence, we apply integration by parts as shown to arrive at the integral strain equation.

$$u = g e^{-\frac{(t - t')}{\lambda}}$$

$$dv = \frac{d\gamma}{dt'}$$

$$\tau(t) = \int_{-\infty}^{t} \left[g e^{-\frac{(t - t')}{\lambda}} \right] \left[\frac{d\gamma}{dt'} \right] dt' = \left[\left(g e^{-\frac{(t - t')}{\lambda}} \right) \gamma \right]_{-\infty}^{t} - \int_{-\infty}^{t} \gamma \left[\frac{g}{\lambda} e^{-\frac{(t - t')}{\lambda}} \right] dt'$$

For the first term $\left[\left(g e^{-\frac{(t - t')}{\lambda}} \right) \gamma \right]_{-\infty}^{t}$, we know from earlier analysis that at current time t, strain γ is zero. This means when $t' = t$, $\gamma = 0$. While at $t' = -\infty$, the exponential term goes to zero. Therefore, the first term goes to zero, and our expression for shear stress simplifies to only the second term.

$$\tau(t) = 0 - \int_{-\infty}^{t} \gamma \left[\frac{g}{\lambda} e^{-\frac{(t - t')}{\lambda}} \right] dt' = - \int_{-\infty}^{t} \gamma \left[\frac{g}{\lambda} e^{-\frac{(t - t')}{\lambda}} \right] dt'$$

We can use the same mathematical "trick" as in part (c) by defining a new time variable s where $s = t - t'$ such that when $t' = t$, $s = 0$ and when $t' = -\infty$, $s = \infty$. Also, $ds = -dt'$.

$$\tau(s) = \int_{\infty}^{0} \gamma \left[\frac{g}{\lambda} e^{-\frac{s}{\lambda}} \right] ds$$

To solve this integral, we recall again the two distinct phases for our problem, the steady shear period when strain rate is constant and the subsequent relaxation period when strain rate is zero. From current time $t' = t$ (or $s = 0$) moving backward in time to $t' = 0$ or ($s = t$), we have $\dot\gamma = 0$. Then moving further backward in time from $t' = 0$ or ($s = t$) to $t' = -\infty$ (or $s = \infty$), we have constant strain rate $\dot\gamma = \dot\gamma_0$ which can be "taken out" of the integral. Therefore, our expression becomes

$$\tau(s) = \int_{\infty}^{t} \gamma \left[\frac{g}{\lambda} e^{-\frac{s}{\lambda}} \right] ds + \int_{t}^{0} \gamma \left[\frac{g}{\lambda} e^{-\frac{s}{\lambda}} \right] ds$$

We know that at current time t, strain is zero. We also know that strain rate is zero from $s = t$ to $s = 0$; therefore the second term goes to zero, and we are left with the term describing the period of constant strain rate.

$$\tau(s) = \int_{\infty}^{t} \gamma \left[\frac{g}{\lambda} e^{-\frac{s}{\lambda}} \right] ds$$

We can express strain as a straight-line function via observation, from our earlier plot and analysis.

$$\gamma = -\dot{\gamma}_0(s-t) = -\dot{\gamma}_0 s + \dot{\gamma}_0 t$$

$$\tau(s) = \int_\infty^t (-\dot{\gamma}_0 s + \dot{\gamma}_0 t)\left[\frac{g}{\lambda}e^{-\frac{s}{\lambda}}\right]ds = \frac{\dot{\gamma}_0 g}{\lambda}\int_\infty^t (t-s)\left[e^{-\frac{s}{\lambda}}\right]ds$$

Applying the method of integration by parts again, whereby

$$u = t - s$$

$$dv = e^{-\frac{s}{\lambda}}$$

$$\frac{\dot{\gamma}_0 g}{\lambda}\int_\infty^t (t-s)\left[e^{-\frac{s}{\lambda}}\right]ds = \frac{\dot{\gamma}_0 g}{\lambda}\left\{\left[(t-s)(-\lambda)e^{-\frac{s}{\lambda}}\right]_\infty^t - \int_\infty^t (-\lambda)(-1)e^{-\frac{s}{\lambda}}ds\right\}$$

$$= \frac{\dot{\gamma}_0 g}{\lambda}\left\{0 + \left[\lambda^2 e^{-\frac{s}{\lambda}}\right]_\infty^t\right\} = \frac{\dot{\gamma}_0 g}{\lambda}\left(\lambda^2 e^{-\frac{t}{\lambda}}\right)$$

$$\tau = \dot{\gamma}_0 g \lambda e^{-\frac{t}{\lambda}} = \eta\dot{\gamma}_0 e^{-\frac{t}{\lambda}} = \tau_0 e^{-\frac{t}{\lambda}}$$

We have thus shown the same result using the integral strain equation.

Problem 9

(a) **Explain what is meant by a viscoelastic fluid, and comment on their industrial applications with suitable examples.**
(b) **Briefly describe experimental methods used to study viscoelastic behavior and the appropriate models to apply to each of these methods. [Hint: You may consider the Maxwell and Voigt elements.]**

Solution 9

Worked Solution

(a) A viscoelastic fluid is a material that exhibits both viscous and elastic properties, and examples of such fluids include polymer melts. Viscoelastic response is useful in the fields of polymer science, polymer engineering, and composites engineering, since the mechanical response of polymers or polymer composites can be associated with the material's structure and chemistry. Specialized polymer materials with desired performance properties can thus be designed for niche applications. In addition, processing issues can be better resolved with an understanding of rheological characteristics and complexities of polymers, thereby allowing enhanced quality control in industrial plants.

Many mathematical models for linear viscoelastic materials have been developed as a starting point to approximate polymer behavior. Although these models can be applied to both solids and liquids, they are often used to model polymer melts (liquid).

Some commercial examples of viscoelastic materials include biotechnological products such as food additives (e.g., Xanthan gum); condiments such as sauces, tomato ketchup, and custard; as well as personal consumer goods such as body lotions, shampoos, and shaving creams. In the aerospace, automotive, and defense industries, an understanding of viscoelastic polymers also helps support applications in fuel, machine parts, and explosives.

(b) Rheometers are often used experimentally in the study of viscoelasticity as they are able to maintain constant strain or stress. There are two main types of experiments one can conduct:

1. Controlled strain experiment – Apply constant strain and measure shear stress.
2. Controlled stress experiment – Apply constant stress and measure strain.

The applied strain (or stress) in rheometers can be steady (e.g., constant value maintained over time), oscillatory, or in steps (e.g., delta step input). The response of the fluid can then be measured, and the data obtained can be fit to analytical/mathematical models. Some simple models include the Maxwell model and the Voigt model for linear viscoelastic fluids. The Maxwell model is often well used to model stress relaxation of viscoelastic materials (e.g., Silly Putty or warm tar) under an applied strain, while the Voigt model is often used to study the deformation response from measured values of strain for viscoelastic materials (e.g., polymers, wood or rubbers) under an applied stress.

Maxwell Model

Spring (1)   Dashpot (2)

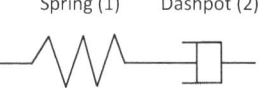

The Maxwell model consists of a linear spring (i.e., constant value for elastic constant g) labelled "1" above and linear dashpot (i.e., constant value for viscosity η) labelled "2" above connected in series. The spring represents the elastic component of the fluid and obeys Hooke's law, while the dashpot represents the viscous component like a Newtonian damper. Under this model, there are two key points to note by virtue of the series connection:

1. Stress continuity – $\tau_{total} = \tau_1 = \tau_2$.
2. Strain additivity – $\gamma_{total} = \gamma_1 + \gamma_2$

Due to stress continuity, it is often preferred to use the Maxwell model for a controlled strain experiment whereby stress values are measured. The constitutive equation for the Maxwell model can be derived as shown below and is a useful starting equation as it effectively relates stress to strain.

$$\gamma_{total} = \gamma_1 + \gamma_2$$

$$\dot{\gamma}_{total} = \dot{\gamma}_1 + \dot{\gamma}_2 = \frac{\dot{\tau}_{total}}{g_1} + \frac{\tau_{total}}{\eta_2}$$

$$g_1\dot{\gamma}_{total} = \dot{\tau}_{total} + g_1\frac{\tau_{total}}{\eta_2}$$

We define relaxation time of the viscoelastic fluid to be $\lambda = \eta_2/g_1$, therefore

$$g_1\dot{\gamma}_{total} = \dot{\tau}_{total} + \frac{\tau_{total}}{\lambda}$$

$$g\frac{d\gamma}{dt} = \frac{d\tau}{dt} + \frac{\tau}{\lambda}$$

Voigt Model

For a controlled stress experiment where strain is measured, we can apply the Voigt model, which is described below.

Spring (1)

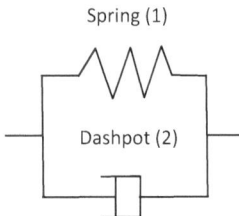

Dashpot (2)

The Voigt element consists of a linear spring and dashpot connected in parallel. By virtue of the parallel connection, there are two key points to note as follows:

1. Strain continuity
2. Stress additivity

Due to strain continuity, it is often preferred to use the Voigt model for a controlled stress experiment. Similarly, the constitutive equation for the Voigt model can be derived as shown below.

$$\gamma_{total} = \gamma_1 = \gamma_2$$

$$\dot{\gamma}_{total} = \dot{\gamma}_1 = \dot{\gamma}_2$$

$$\tau_{total} = \tau_1 + \tau_2$$

$$\tau_{total} = g_1\gamma_1 + \eta_2\dot{\gamma}_2$$

$$\tau = g\gamma + \eta\frac{d\gamma}{dt}$$

The Voigt model can be used to study deformation of viscoelastic materials under a constant applied stress. The material deforms at a decreasing rate and asymptotically approaches a steady-state value for strain. This phenomenon is also known as "creep," and the Voigt model is commonly used to model this behavior.

Problem 10

Linear viscoelastic behavior can be modelled using the Maxwell integral strain rate equation given below where $\tau(t)$ denotes shear stress at a current time t, $t^{'}$ denotes a past time variable, g denotes the elastic constant for the elastic component of the material, $\dot{\gamma}$ denotes strain rate, and λ denotes relaxation time of the material.

$$\tau(t) = \int_{t^{'}=-\infty}^{t^{'}=t} g\dot{\gamma}\left(t^{'}\right)e^{-\frac{(t-t^{'})}{\lambda}}dt^{'}$$

(a) **In a controlled strain experiment, the applied strain rate is as shown in the plot below. Determine the resultant stress response, and explain using a suitable plot, how the fluid would behave if it was completely inelastic.**

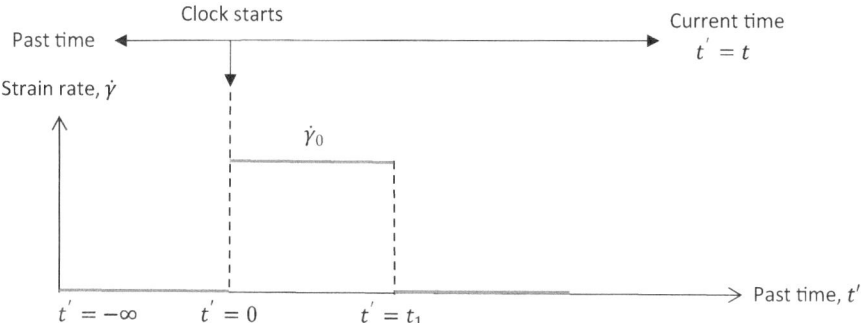

(b) **Discuss the limitations of the Maxwell model, and suggest ways to overcome these limitations when modelling non-linear viscoelastic behavior (e.g., shear-thinning).**

Solution 10

Worked Solution

(a) Let us dissect the strain rate profile into three distinct time periods as indicated below in red, and apply the Maxwell equation to each period accordingly.

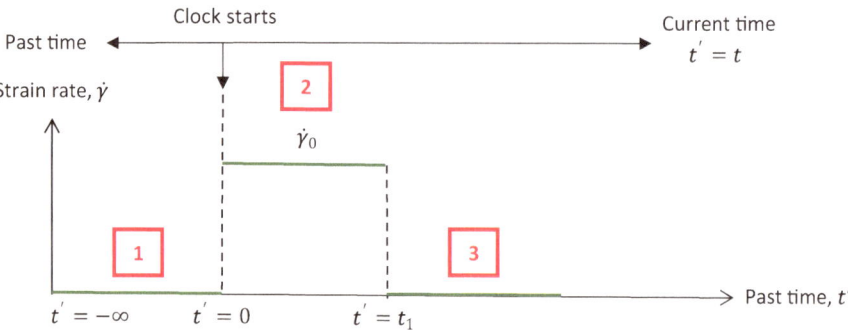

Time Period 1: Historical Time with Zero Strain Rate

During the period from $t' = -\infty$ to $t' = 0$, the input strain rate is zero, i.e., $\dot{\gamma}(t') = 0$. Therefore, it follows from the integral strain rate equation that shear stress is also zero. Note that by definition, the left-hand side of the expression below refers to shear stress $\tau(t)$ at an arbitrary *current* time t whereby $t = 0$ is the point at which the clock starts.

$$\tau(t) = \int_{t'=-\infty}^{t'=t} g\dot{\gamma}(t')e^{-\frac{(t-t')}{\lambda}}dt' = \int_{-\infty}^{t} g(0)e^{-\frac{(t-t')}{\lambda}}dt' = 0$$

Time Period 2: Current Time with Constant Applied Strain Rate

During the period from $t' = 0$ to $t' = t_1$, there is a constant applied strain rate of $\dot{\gamma}_0$. At this point, we note that shear stress at current time t is also affected by strain rate history. Hence, we need to consider prior periods of strain rate when considering this time period. Mathematically, this is equivalent to determining the sum of integrals with limits of integration as shown below.

$$\tau(t) = \int_{t'=-\infty}^{t'=0} g\dot{\gamma}(t')e^{-\frac{(t-t')}{\lambda}}dt' + \int_{t'=0}^{t'=t} g\dot{\gamma}(t')e^{-\frac{(t-t')}{\lambda}}dt' = 0 + \int_{0}^{t} g\dot{\gamma}(t')e^{-\frac{(t-t')}{\lambda}}dt'$$

Substituting the value for strain rate into the integral equation, we obtain an exponential function that describes the shape of the plot of stress response. This type of response is also referred to as "stress growth."

$$\tau(t) = \dot{\gamma}_0 \int_{0}^{t} ge^{-\frac{(t-t')}{\lambda}}dt' = \dot{\gamma}_0 \int_{0}^{t} ge^{-\frac{t}{\lambda}}e^{\frac{t'}{\lambda}}dt'$$

$$\tau(t) = \dot{\gamma}_0 ge^{-\frac{t}{\lambda}}\left[\lambda e^{\frac{t'}{\lambda}}\right]_{0}^{t} = \lambda\dot{\gamma}_0 ge^{-\frac{t}{\lambda}}\left[e^{\frac{t}{\lambda}} - 1\right] = \lambda\dot{\gamma}_0 g\left(1 - e^{-\frac{t}{\lambda}}\right)$$

Note that by definition, relaxation time λ is equivalent to

$$\lambda = \frac{\eta}{g}$$

$$\tau(t) = \eta\dot{\gamma}_0\left(1 - e^{-\frac{t}{\lambda}}\right)$$

We may also observe from the above expression that relaxation time affects the rate of exponential growth for stress in this time period.

For a fully viscous fluid that is nonelastic, $\lambda = 0$, the expression simplifies to the following. Assuming a Newtonian fluid where viscosity η is constant, we will have a straight horizontal line for shear stress at a constant value of $\tau = \eta\dot{\gamma}_0$, instead of an exponentially increasing profile for a linear viscoelastic fluid.

$$\tau(t) = \eta\dot{\gamma}_0\left(1 - e^{-\frac{t}{0}}\right) = \eta\dot{\gamma}_0(1 - e^{-\infty}) = \eta\dot{\gamma}_0(1 - 0) = \eta\dot{\gamma}_0$$

As the relaxation time of a fluid increases, the more elastic it is, and the slower it takes for shear stress to exponentially grow. At the other extreme case where we have a fully elastic fluid with no viscous components, $\lambda = \infty$ and our expression simplifies to the following where shear stress becomes zero.

$$\tau(t) = \eta\dot{\gamma}_0\left(1 - e^{-\frac{t}{\alpha}}\right) = 0$$

Time Period 3: Current Time with Zero Strain Rate

During the period from $t' = t_1$ to $t' = t$, the applied strain rate returns to zero. Similar to before, we need to consider all prior periods for strain rate in order to plot the stress response in this time period.

$$\tau(t) = \int_{t'=-\infty}^{t'=0} g\dot{\gamma}(t')e^{-\frac{(t-t')}{\lambda}}dt' + \int_{t'=0}^{t'=t_1} g\dot{\gamma}(t')e^{-\frac{(t-t')}{\lambda}}dt' + \int_{t'=t_1}^{t'=t} g\dot{\gamma}(t')e^{-\frac{(t-t')}{\lambda}}dt'$$

$$\tau(t) = 0 + \dot{\gamma}_0\int_0^{t_1} ge^{-\frac{(t-t')}{\lambda}}dt' + 0 = \dot{\gamma}_0\int_0^{t_1} ge^{-\frac{t}{\lambda}}e^{\frac{t'}{\lambda}}dt'$$

$$\tau(t) = \dot{\gamma}_0 ge^{-\frac{t}{\lambda}}\left[\lambda e^{\frac{t'}{\lambda}}\right]_0^{t_1} = \lambda\dot{\gamma}_0 ge^{-\frac{t}{\lambda}}\left[e^{\frac{t_1}{\lambda}} - 1\right] = \left[\eta\dot{\gamma}_0\left(e^{\frac{t_1}{\lambda}} - 1\right)\right]e^{-\frac{t}{\lambda}}$$

Note that the quantity $\left[\eta\dot{\gamma}_0\left(e^{\frac{t_1}{\lambda}} - 1\right)\right]$ is simply a constant as t_1 is a fixed value. Therefore, we obtain an exponential decay response for shear stress as a function of current time t.

Putting together our results above, we have the following plot of shear stress τ.

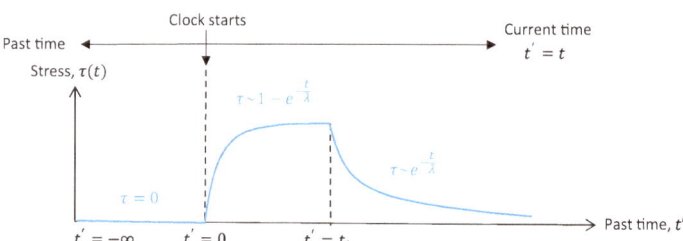

The following plot shows how stress responses for different cases of inelastic fluids.

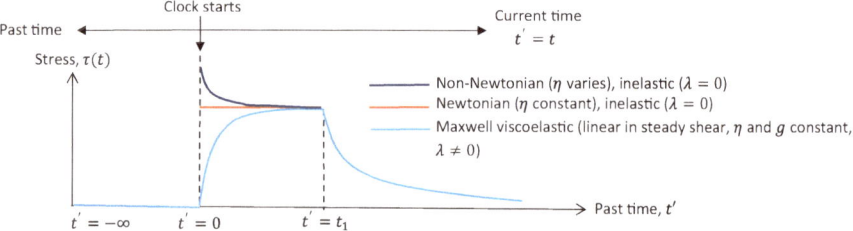

(b) The Maxwell element models stress growth and stress relaxation of viscoelastic materials and predicts a linear response under steady shear. For viscoelastic materials, relaxation time is non-zero; hence they can be represented using a linear Newtonian dashpot (constant viscosity η) and a linear Hookean spring (constant elastic constant g) connected in series.

However, there are some limitations of the Maxwell element as listed below:

1. The Maxwell model predicts Newtonian behavior well and is less suitable for non-Newtonian materials when viscosity η is not a constant (e.g., shear-thinning fluids).
2. The Maxwell model also fails when a material encompasses multiple relaxation times, giving rise to a non-linear viscoelastic response.

In order to overcome the first limitation, we can modify the Maxwell constitutive equation (in terms of past strain) by adding a non-linear factor (i.e., $e^{-k|\gamma|}$ as shown below) to account for the additional strain dependence that non-Newtonian fluids have (e.g., shear-thinning fluids).

$$\tau(t) = -\int_{-\infty}^{t} \frac{g}{\lambda} e^{-\frac{(t-t')}{\lambda}} \dot{\gamma} dt' \rightarrow -\int_{-\infty}^{t} \frac{g}{\lambda} e^{-\frac{(t-t')}{\lambda}} e^{-k|\gamma|} \dot{\gamma} dt'$$

Note that the value of k ranges from 0 to 1. The lower limit $k = 0$ returns a linear Maxwell model. As the value of k increases, the strain dependence and hence degree of non-linearity of the fluid increase correspondingly.

In order to overcome the second limitation, we can couple multiple Maxwell elements in parallel as shown below (using an example of two coupled elements, $n = 2$).

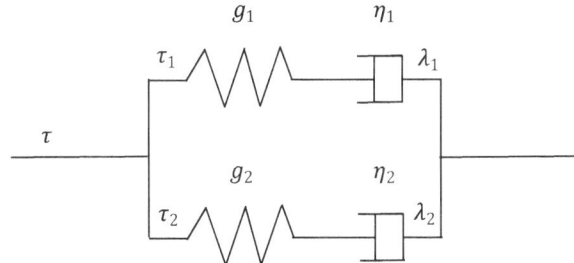

We may now derive the total shear stress of the fluid as the sum of stresses in each parallel branch of the Maxwell element.

$$\tau = \sum_{i=1}^{n} \tau_i = \tau_1 + \tau_2$$

$$g = \sum_{i=1}^{n} g_i = g_1 + g_2$$

$$\eta = \sum_{i=1}^{n} \eta_i = \eta_1 + \eta_2 = g_1\lambda_1 + g_2\lambda_2$$

Note that the strain in each branch is equivalent to each other and is also the same as the total strain, i.e., there is strain continuity in a parallel connection.

$$\gamma = \gamma_1 = \gamma_2$$

We may now put together our earlier findings as developed above, in considering the modified form of the constitutive equation (in terms of past strain) that helps us overcome the second limitation of multiple relaxation times.

$$\tau(t) = -\int_{-\infty}^{t} \frac{g}{\lambda} e^{-\frac{(t-t')}{\lambda}} \gamma dt' \rightarrow -\int_{-\infty}^{t} \sum_{i=1}^{n} \frac{g_i}{\lambda_i} e^{-\frac{(t-t')}{\lambda_i}} \gamma dt'$$

This equation now more appropriately models fluids with multiple relaxation times as it includes the summation term that accounts for the different relaxation times in each parallel branch.

Problem 11

The following form of the Maxwell integral strain equation is used to model shear stress at current time t for non-linear viscoelastic materials which may also comprise of multiple relaxation times (represented by λ_i). Non-linearity is accounted for in the term $e^{-k|\gamma|}$ which represents a varying viscosity η with strain. In the equation below, g denotes elastic constant, while k is a constant with a value between 0 and 1.

$$\tau(t) = -\int_{-\infty}^{t} \sum_{i=1}^{n} \frac{g_i}{\lambda_i} e^{-k|\gamma|} e^{-\frac{(t-t')}{\lambda_i}} \gamma dt'$$

(a) **Under steady shear at a constant strain rate $\dot{\gamma}(t') = \dot{\gamma}_0$, show that the stress response can be represented by the following expression:**

$$\tau = \dot{\gamma}_0 \sum_{i=1}^{n} \frac{\eta_i}{(k\dot{\gamma}_0\lambda_i + 1)^2}$$

(b) **Using the expression in part (a), plot the stress responses for the following cases for a viscoelastic material:**

(i) **Single relaxation time with constant viscosity**
(ii) **Single relaxation time with shear-thinning**
(iii) **Multiple relaxation times with constant viscosity**
(iv) **Multiple relaxation times with shear-thinning**

Solution 11

Worked Solution

(a) We are told that there is a constant applied strain rate $\dot{\gamma}$, i.e., "steady shear" as shown in the plot below.

We can derive an expression for strain γ, by measuring it backward from current time t to a past time t'.

$$\gamma(t, t') = \int_t^{t'} \dot{\gamma}dt' = \dot{\gamma}_0 \int_t^{t'} dt' = \dot{\gamma}_0(t' - t) = -\dot{\gamma}_0(t - t')$$

We can derive the strain profile (as shown in purple below) as we note that the above equation describes a simple straight line. To simplify subsequent mathematical steps, we may also define a new time variable s whereby $s = t - t'$. The corresponding values of s are also indicated in the plot.

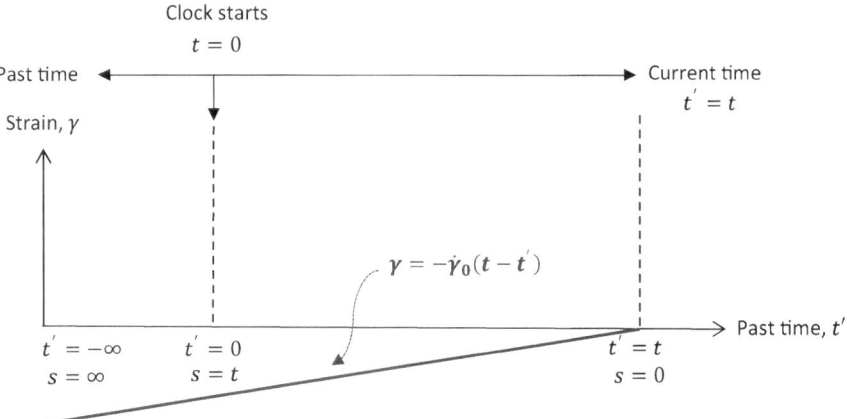

Throughout the period from current time t (when $s = 0$) to an arbitrary time in history (when $0 < s < \infty$), strain can be expressed as follows.

$$\gamma = -\dot{\gamma}_0 s$$

We proceed to substitute the above expression into the Maxwell integral strain equation as shown in purple in the equation below.

$$\tau(t) = -\int_{-\infty}^{t} \sum_{i=1}^{n} \frac{g_i}{\lambda_i} e^{-k|-\dot{\gamma}_0 s|} e^{-\frac{(t-t')}{\lambda_i}} (-\dot{\gamma}_0 s)\, dt' = -\int_{-\infty}^{t} \sum_{i=1}^{n} \frac{g_i}{\lambda_i} e^{-k\dot{\gamma}_0 s} e^{-\frac{(t-t')}{\lambda_i}} (-\dot{\gamma}_0 s)\, dt'$$

Since $s = t - t'$, it follows that $ds = -dt'$. We also know the limits for the above integral with respect to s, $s = \infty$ when $t' = -\infty$ and $s = 0$ when $t' = t$; therefore we arrive at the following expression.

$$\tau(s) = -\int_{s=\infty}^{s=0} \sum_{i=1}^{n} \frac{g_i}{\lambda_i} e^{-k\dot{\gamma}_0 s} e^{-\frac{s}{\lambda_i}} (\dot{\gamma}_0 s)\, ds$$

Now the integral is uniformly expressed in terms of new variable s. Let us combine the exponential terms into a single term, and simplify the expression by grouping parameters together into a single new parameter denoted β_i as shown below.

$$\beta_i = k\dot{\gamma}_0 + \frac{1}{\lambda_i}$$

$$\tau(s) = -\int_\infty^0 \sum_{i=1}^n \frac{g_i}{\lambda_i} e^{-\left(k\dot{\gamma}_0 + \frac{1}{\lambda_i}\right)s} (\dot{\gamma}_0 s) ds = -\int_\infty^0 \sum_{i=1}^n \frac{g_i}{\lambda_i} e^{-\beta_i s} (\dot{\gamma}_0 s) ds$$

In order to solve this integral, we can use the method of integration by parts where the following holds.

$$\int u \, dv = uv - \int v \, du$$

$$u = s \rightarrow du = ds$$

$$dv = -\sum_{i=1}^n \frac{g_i}{\lambda_i} e^{-\beta_i s} (\dot{\gamma}_0) ds \rightarrow v = \sum_{i=1}^n \frac{g_i}{\lambda_i \beta_i} e^{-\beta_i s} (\dot{\gamma}_0)$$

$$\tau(s) = \left[\left(\sum_{i=1}^n \frac{g_i}{\lambda_i \beta_i} e^{-\beta_i s} (\dot{\gamma}_0) \right) s \right]_\infty^0 - \int_\infty^0 \left[\sum_{i=1}^n \frac{g_i}{\lambda_i \beta_i} e^{-\beta_i s} (\dot{\gamma}_0) \right] ds$$

$$= 0 - \left[\sum_{i=1}^n -\frac{g_i}{\lambda_i \beta_i^2} e^{-\beta_i s} (\dot{\gamma}_0) \right]_\infty^0$$

$$\tau(s) = \left[\sum_{i=1}^n \frac{g_i}{\lambda_i \beta_i^2} e^{-\beta_i s} (\dot{\gamma}_0) \right]_\infty^0 = \sum_{i=1}^n \frac{g_i}{\lambda_i \beta_i^2} (\dot{\gamma}_0) - 0 = \sum_{i=1}^n \frac{g_i \dot{\gamma}_0}{\lambda_i \beta_i^2}$$

We can now re-express our terms in terms of our original parameters and variables.

$$\tau = \sum_{i=1}^n \frac{g_i \dot{\gamma}_0}{\lambda_i \left(k\dot{\gamma}_0 + \frac{1}{\lambda_i}\right)^2} = \sum_{i=1}^n \frac{\lambda_i g_i \dot{\gamma}_0}{\lambda_i^2 \left(k\dot{\gamma}_0 + \frac{1}{\lambda_i}\right)^2}$$

Note that by definition, relaxation time λ can be expressed as such

$$\lambda = \frac{\eta}{g}$$

Therefore, we simplify our stress equation into the given form as shown below.

$$\tau = \sum_{i=1}^n \frac{\eta_i \dot{\gamma}_0}{(k\dot{\gamma}_0 \lambda_i + 1)^2} = \dot{\gamma}_0 \sum_{i=1}^n \frac{\eta_i}{(k\dot{\gamma}_0 \lambda_i + 1)^2}$$

We have derived the above expression which describes stress response under steady shear for a viscoelastic material that comprises of multiple relaxation times (up to n numbers of relaxation times in the summation). Note that shear stress τ

depends on the value of applied strain rate $\dot{\gamma}_0$. We can express the stress equation in the form as follows, in order to examine the fluid's apparent viscosity η_{app}.

$$\tau = \dot{\gamma}_0 \sum_{i=1}^{n} \frac{\eta_i}{(k\dot{\gamma}_0\lambda_i + 1)^2} = \dot{\gamma}_0\eta_{app}$$

We note that the apparent viscosity, η_{app}, for this viscoelastic fluid may not be a constant like in the case of a Newtonian fluid (the case when $k = 0$). η_{app} is now variable and dependent on properties characterized by parameters such as λ_i, k, and η_i as shown below.

$$\eta_{app} = \sum_{i=1}^{n} \frac{\eta_i}{(k\dot{\gamma}_0\lambda_i + 1)^2}$$

Therefore, the expression above may also be used to describe non-Newtonian behavior such as shear-thinning.

(b) (i) For a single relaxation time ($n = 1$) with constant viscosity ($k = 0$), we have the basic Newtonian fluid. The expression is reduced to the following.

$$\tau = \dot{\gamma}_0 \sum_{i=1}^{n=1} \frac{\eta_i}{((0)\dot{\gamma}_0\lambda_i + 1)^2} = \eta\dot{\gamma}_0$$

The plot is therefore linear as expected for Newtonian behavior. It makes sense that with increasing viscosity, there is greater stress for a given strain rate.

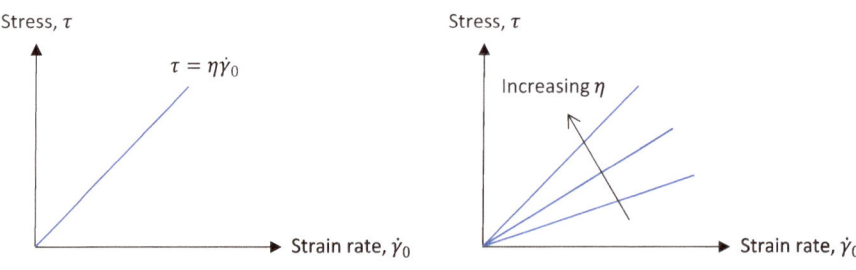

(b) (ii) Now let us consider a Non-Newtonian viscoelastic material with shear-thinning characteristics. This means that viscosity is no longer constant but becomes smaller at higher strain rates. (Note that fluids may also be shear-thickening, which means a larger viscosity at higher strain rates.)

In the case of a shear-thinning fluid, we have a single relaxation time; hence the summation is not necessary (in other words, $n = 1$ for the summation). However, k is now a non-zero value. The result is a non-linear function for stress as shown below.

We can plot this function using Excel by setting arbitrary values for the parameters η, k, and λ, for a defined range of strain rates.

$$\tau = \dot{\gamma}_0 \sum_{i=1}^{n=1} \frac{\eta_i}{(k\dot{\gamma}_0\lambda_i + 1)^2} = \dot{\gamma}_0 \frac{\eta}{(k\dot{\gamma}_0\lambda + 1)^2}$$

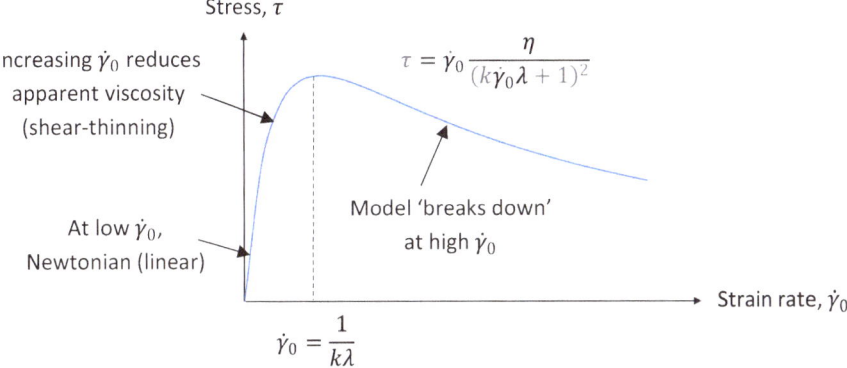

Notice that the plot is non-linear with a maximum point. We can observe that at very low strain rates whereby $k\dot{\gamma}_0\lambda \ll 1$, the fluid tends to become more Newtonian-like, whereby the graph tends to a linear plot.

$$\tau = \dot{\gamma}_0 \frac{\eta}{(k\dot{\gamma}_0\lambda + 1)^2} \cong \dot{\gamma}_0\eta$$

The shear-thinning characteristic of the fluid is well captured by this model at lower values of strain rate (before the maximum point) as it shows non-linearity which represents a falling apparent viscosity as strain rate increases. However, at high strain rates, we start to observe that two solutions for strain rate exist for a single value of stress. This is not physically meaningful in most cases where we have stable materials. Therefore, this mathematical model is best applied for lower values of strain rates before the maxima in the plot.

Mathematically, we can find that the maximum point occurs at $\dot{\gamma}_0 = \frac{1}{k\lambda}$ by setting the first derivative to zero as shown below, where the differentiation is done using product rule.

$$\frac{d\tau}{d\dot{\gamma}} = 0 = \frac{-2k\lambda\eta\dot{\gamma}_0}{(k\dot{\gamma}_0\lambda + 1)^3} + \frac{\eta}{(k\dot{\gamma}_0\lambda + 1)^2}$$

$$\frac{2k\lambda\dot{\gamma}_0}{k\dot{\gamma}_0\lambda + 1} = 1$$

$$\dot{\gamma}_0\big|_{\tau_{\max}} = \frac{1}{k\lambda}$$

From the above expression, we observe the dependency of maximum stress $\tau_{\max}$ on the damping parameter k as well as on relaxation time λ. As damping factor k increases, the maximum point is shifted left (occurs at smaller strain rate). Similarly, as relaxation time λ increases, the maximum point is shifted left. Furthermore, the values of stress are broadly reduced as k increases. This makes sense since the decaying effect of the exponential term ($e^{-k|\gamma|}$) on shear stress is enhanced as k increases. The values of stress are also broadly reduced as λ increases, similar to the effect that k has on the function for stress τ.

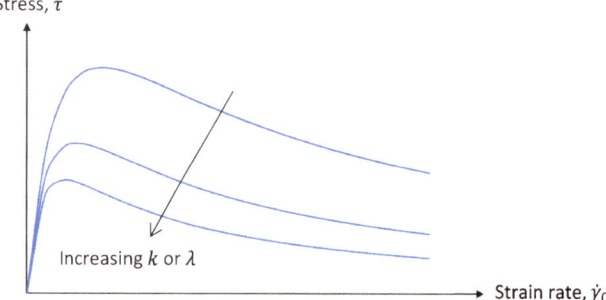

We can study the dependence of viscosity on shear stress for such fluids. As viscosity increases, we note that the values of stress broadly increase for a given strain rate, which also makes physical sense.

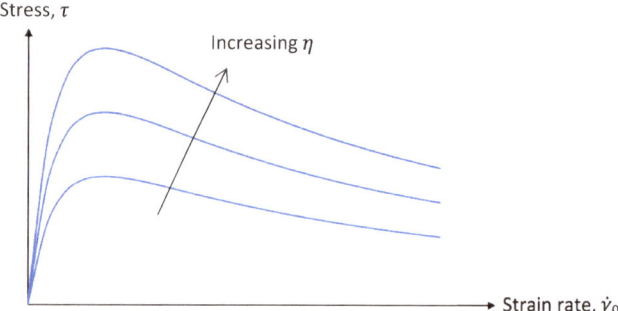

(b) (iii) Let us now explore the case where we have multiple relaxation times but a constant Newtonian viscosity ($k = 0$). Our expression for stress can be simplified as follows. We note that the expression returns back to the form of a Newtonian fluid. Therefore, we expect our plot to be linear like in the case of (b) (i).

However this time, the value of apparent viscosity is the sum of all component viscosities, whereby $\eta_i = \lambda_i g_i$. Assuming we have a material that can be modelled by two component relaxation times, i.e., $n = 2$ for the summation, we obtain the following plots.

$$\tau = \dot{\gamma}_0 \sum_{i=1}^{n} \frac{\eta_i}{((0)\dot{\gamma}_0\lambda_i + 1)^2} \cong \dot{\gamma}_0 \sum_{i=1}^{n} \eta_i = \dot{\gamma}_0\eta_{total}$$

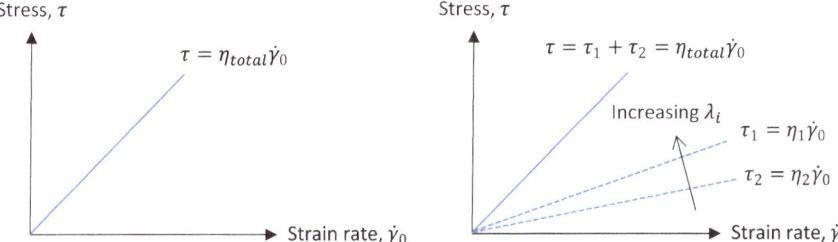

(b) (iv) We now have the most complex scenario where the fluid is both shear-thinning ($k \neq 0$) and represented by multiple relaxation times ($n > 1$ for the summation). To plot this graph, we can sum separate graphs with each representing a specific relaxation time. In this example, we assume that our fluid may be modelled by three constituent relaxation times λ_1, λ_2, and λ_3 where $\lambda_3 > \lambda_2 > \lambda_1$. We observe that the shape of the graph for each component i in the summation is the same as that in part (b) (ii). The summation effectively "corrects" the right-hand side of the graph in part (b) (ii) by removing the previously "physically unrealistic" portion (where the model broke down as the graph "dipped" at high strain rates) by raising the values of stress at higher strain rates such that there are no longer two values of strain rates (i.e., two solutions as in a quadratic function) for a single value of stress. Hence this model is the most robust out of the four cases explored in this problem when modelling a real fluid with shear-thinning properties (e.g., power law fluids).

$$\tau = \dot{\gamma}_0 \sum_{i=1}^{n} \frac{\eta_i}{(k\dot{\gamma}_0\lambda_i + 1)^2} = \dot{\gamma}_0 \frac{\eta_1}{(k\dot{\gamma}_0\lambda_1 + 1)^2} + \dot{\gamma}_0 \frac{\eta_2}{(k\dot{\gamma}_0\lambda_2 + 1)^2} + \dot{\gamma}_0 \frac{\eta_3}{(k\dot{\gamma}_0\lambda_3 + 1)^2}$$

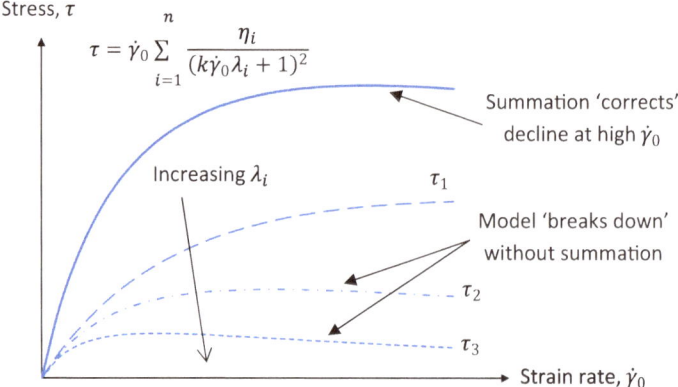

Problem 12

In the processing industry, it is useful to be familiar with the viscosities of fluids and how they behave under different operating conditions. As such, various mathematical models have been developed to help characterize and predict the physical behaviors of molten polymers with viscoelastic properties.

(a) Describe how viscoelastic polymer melts may be represented using a model comprising of a viscous component and an elastic component. Cite some examples of viscoelastic fluids and the typical range of values for viscosity, elasticity, and relaxation times for these fluids.

(b) Determine the magnitudes of both elasticity and viscosity for the integral form of the Maxwell model as given below, whereby $\gamma(t, t')$ denotes strain between current time t and a past time t'.

$$\tau(t) = -\int_{-\infty}^{t} 1020 e^{-(t-t')/10} \gamma(t, t') dt'$$

(c) The following expression may be used to predict the stress response of a power law fluid. Explain what is meant by a power law fluid and how it differs from a Newtonian fluid, using suitable plots that portray stress and viscosity behaviors.

$$\tau = k \dot{\gamma}^n$$

Solution 12

Worked Solution

(a) The properties of polymer melts may be described by mathematical models that break down the analysis of their viscoelasticity into a purely elastic component and a purely viscous (inelastic) component. For small deformations, the response of the system can be approximated as linear; therefore we can mathematically model easily using the addition principle, whereby the overall response of the linear viscoelastic material is the sum of two individual responses (contributed by the elastic and viscous components, respectively) to a small deformation.

Elastic Component

The elastic component uses the concept of a spring, whereby the polymer chain deforms in a way similar to how a spring extends under an external force. One common example is the typical rubber which is essentially a polymer that stretches under an applied force. A point to note is that elasticity is not a property exclusive to polymer liquids. Almost all liquids exhibit elastic behavior at sufficiently small time scales.

The elastic component is best explained using thermodynamics, particularly entropy. Entropy is defined as the degree of disorder or randomness in a system. The second law of thermodynamics states that the total entropy of a system either increases or remains constant, i.e., it cannot decrease. Entropy is zero in a reversible process and increases in an irreversible process.

A polymer melt is in fact a macromolecule comprising a large network of cross-linked polymer chains. The thermodynamically preferred state of a polymer chain is a configuration that minimizes the distance between its free ends as this maximizes entropy. However when an external force is applied, the polymer deforms and stretches beyond its equilibrium end-to-end distance, causing a reduction in entropy. This gives rise to an elastic force that opposes this strain. These "springs" may be restored to their equilibrium state using this elastic energy that was "stored" during the prior deformation process.

It is important to note that elasticity is defined as reversible deformation under an applied stress. Assuming a linear elastic force, we may then use Hooke's law to relate the extension length to the applied stress. Elasticity is dependent on temperature (generally less elastic at higher temperature) as well as chain length (longer chains with higher molecular weights are generally more elastic).

We use the elastic constant g [Pa] to characterize elasticity and quantify the amount of "recoverable" energy stored by the deformation. It is defined as follows, where τ denotes shear stress and λ denotes relaxation time.

$$g = \frac{\tau}{\lambda}$$

The value of g for most polymer melts lies in the range of $10^{-1} - 10^4$ Pa, and this is much lower than that of polymeric solids which typically exceed 10^{10} Pa.

Viscous Component

Separately, we use a purely inelastic element to characterize the polymer's viscous property which represents the dissipative reaction to deformation. Viscosity is a measure of a fluid's resistance to flow due to internal friction at the molecular level. We use η [Pa. s] to measure viscosity where the higher its value, the "thicker" the fluid and greater the resistance to gradual deformation under a shear or tensile stress. It follows that a higher viscosity means a greater ability to experience drag force. η may be defined as follows in terms of shear stress τ and strain (or shear) rate $\dot{\gamma}$.

$$\eta = \frac{\tau}{\dot{\gamma}}$$

Viscosity depends on several factors, one of which is polymer size (or molecular weight M). Most commercial polymers are polydisperse, meaning they comprise of a distribution of molecular masses in a given sample. In general, the larger the polymer size, the greater the viscosity. The viscosity of low M fluids (many organic liquids) is typically in the range of 10^{-3} (e.g., water) to 10^{-2} Pa. s, and the fluid flows easily in this range.

As viscosity increases to about 10 Pa. s, we have a "thicker" or "stickier" fluid similar in consistency to glycerol and honey. As we increase viscosity further to around $10^2 - 10^3$ Pa. s, applied pressure is usually necessary to drive fluid flow, and this range applies for high M fluids such as certain polymer melts of moderate chain lengths. Viscosities for large polymer melts (very high M) comprising networks of long cross-linked chains can go up to around 10^5 to 10^6 Pa.s.

Viscosity also depends on temperature according to the following expression which is in the form of an Arrhenius equation consisting of a pre-exponential η_0, activation energy E, and the gas constant R.

$$\eta = \eta_0 e^{\frac{E}{RT}}$$

Knowledge of viscosity is important in designing processes as it is ideal to have operating conditions such that fluids are easily handled. For example, if the behavior of viscosity with respect to temperature was known, one could avoid high-temperature processing whereby the fluid becomes too hot and "runny" (flows too easily) to control. The typical range of viscosity for ideal processability lies between around 10^{-1} Pa. s and 10^2 Pa. s.

	Relaxation time λ [s]	Viscosity η [Pa. s]	Elastic constant g [Pa]
Water, inkjet (printing)	10^{-12}	10^{-3}	10^9
Oil, organic liquids	10^{-9}	10^{-1}	10^8
Honey, glycerol, polymer solutions	1	10	10
Polymer melts (scales with size)	10	Up to about 10^6	Up to about 10^5

Relaxation time λ is defined as follows and typically scales with molecular weight to the order of 3 or 4 for large entangled chains of polymer melts ($\lambda \sim M^{3-4}$):

$$\lambda = \frac{\eta}{g}$$

(b) The general form of the Maxwell equation can be expressed as follows:

$$\tau(t) = -\int_{-\infty}^{t} \frac{g}{\lambda} e^{-(t-t')/\lambda} \gamma(t,t') dt'$$

Therefore, by observation we can determine the required quantities.

$$\tau(t) = -\int_{-\infty}^{t} 1020 e^{-(t-t')/10} \gamma(t,t') dt'$$

$$\frac{g}{\lambda} = 1020, \quad \lambda = 10s$$

The values of elastic constant g and viscosity η can be computed easily.

$$g = 1020(10) = 10200 Pa$$

$$\eta = g\lambda = 10200(10) = 102000 Pas$$

(c) The power law model is commonly used to analyze non-Newtonian, non-linear behavior of fluids such as polymer melts due to the relative ease in the mathematical steps involved. The shear stress τ of a power law fluid can be expressed as follows where strain rate $\dot{\gamma}$ is raised to the nth power and k is a constant with units that depend on the value of n.

$$\tau = k\dot{\gamma}^n$$

In general, we may express the shear stress of any fluid (Newtonian or non-Newtonian) based on the apparent (or observed) viscosity, η_{app} in the expression

$\tau = \eta_{app}\dot{\gamma}$. Another form of the above equation is shown below, and it is useful since it can be represented by a linear plot.

$$\tau = \eta_{app}\dot{\gamma}$$

Equating both expressions for τ, we obtain

$$\eta_{app}\dot{\gamma} = k\dot{\gamma}^{n}$$

$$\eta_{app} = k\dot{\gamma}^{n-1} \qquad (1)$$

Taking logarithms on both sides, we have

$$\ln \eta_{app} = \ln k + (n - 1)\ln\dot{\gamma}$$

A Newtonian fluid exhibits a linear change in shear stress with strain rate, i.e., $n = 1$ as follows where η is a constant and independent of strain rate. Therefore it can be said that a Newtonian fluid responds to the instantaneous strain rate.

$$\tau = k\dot{\gamma}, \ \ k = \eta_{app} = \eta$$

A non-Newtonian fluid may be either shear-thinning (usually lower molecular mass representing most polymer melts of processing grade) or shear-thickening (usually higher molecular mass) where the former means that viscosity reduces as strain rate increases and the latter means that viscosity increases as strain rate increases. Mathematically, we observe from Eq. 1 that shear-thinning power law fluids have $n < 1$ while shear-thickening power law fluids have $n > 1$. An example of a polymer melt well characterized by the power law model is polyethylene, commonly found in plastic bags and plastic bottles. It is a relatively inexpensive material that may exist as linear or branched chains and of high or low density. Another example is polystyrene which is an optically transparent and amorphous polymer commonly found in more rigid food packaging containers and construction materials.

The plots below illustrate the differences between a non-Newtonian power law fluid and a Newtonian fluid.

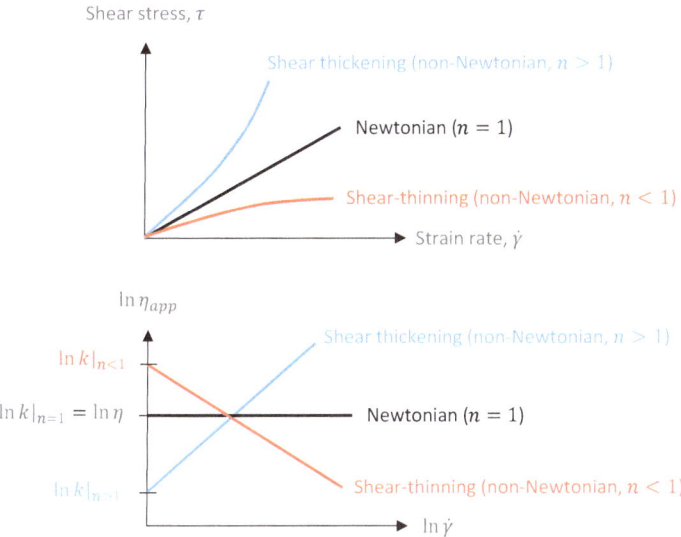

A power law model works best to fit experimental data obtained for a real fluid that shear thins (labelled in orange above) at higher strain rates when the fluid starts to exhibit non-linear dependence of viscosity on strain rate. At lower strain rates, the power law model is less necessary since the power law fluid still behaves like a Newtonian fluid, as shown by the horizontal linear region toward the left in the plot below.

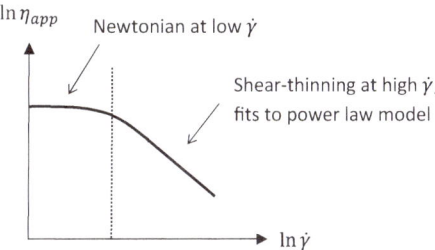

Problem 13

Compare and contrast various models that may be used to describe non-Newtonian shear-thinning fluids, including the power law model, Carreau model, and Cross model in your examples.

Solution 13

Worked Solution

The power law model is widely popular in characterizing non-Newtonian fluids (both shear-thinning and shear-thickening) in process engineering applications, especially in studies on the handling of fluid foods. The general form of its equation is shown below for shear stress τ and apparent viscosity η_{app}, where $\dot{\gamma}$ denotes shear rate (or strain rate) and k and n are constants. For shear-thinning fluids, the exponent $n < 1$.

$$\eta_{app} = k\dot{\gamma}^{n-1}$$

$$\tau = \eta_{app}\dot{\gamma} = k\dot{\gamma}^{n}$$

The power law model is mathematically easy to handle in terms of the relatively simple form of its equation, as well as being just a two-parameter (k and n) model. One should note that it is applicable over a shear rate range of about $10^{1} - 10^{4} s^{-1}$, and this range falls within the operating limits of most commercial viscometers. However, as the power law model is empirical, it should only be used for the intended range of shear rates. At very high or very low shear rates, this model may not be suitable.

The plot below shows the shear stress profile of a power law model.

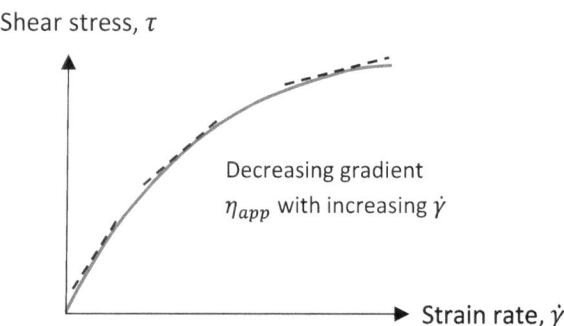

In reality, a single fluid sample may exhibit different types of behaviors at different shear rates. Therefore a model that caters to different behaviors under different conditions would be useful. The Carreau model is one example.

The Carreau model is suitable when the non-Newtonian fluid exhibits two distinct types of behaviors, namely, an Upper Newtonian region at lower shear rates and a power law region at higher shear rates. This model improves upon the power law model by including an extra parameter, η_{0} (also called the "constant zero-shear-rate viscosity"), to take into consideration the Newtonian region. This model can be used to analyze fluids such as blood samples and most commercial polymer melts.

$$\eta_{app} = \frac{\eta_0}{\left[1 + (\beta\dot{\gamma})^2\right]^m}$$

In the above equation, β and m are constant parametric values. Upon inspection of the Carreau equation, we may observe that at low shear rates, we have Newtonian behavior where viscosity is a constant value.

$$\beta\dot{\gamma} \ll 1$$

$$\eta_{app} \cong \frac{\eta_0}{[1]^m} = \eta_0$$

$$\tau = \eta_{app}\dot{\gamma} = \eta_0\dot{\gamma}$$

At high shear rates, we notice that the Carreau equation simplifies to the power law model.

$$\beta\dot{\gamma} \gg 1$$

$$\eta_{app} \cong \frac{\eta_0}{\left[(\beta\dot{\gamma})^2\right]^m} = \eta_0\left(\beta^{-2m}\right)\left(\dot{\gamma}^{-2m}\right)$$

If we define a new parameter n which appears in the exponent term, whereby $n - 1 = -2m$, we get back the form of the power law model where k is a constant such that $k = [\eta_0(\beta^{n-1})]$.

$$\eta_{app} = \left[\eta_0\left(\beta^{n-1}\right)\right]\dot{\gamma}^{n-1} = k\dot{\gamma}^{n-1}$$

$$\tau = \eta_{app}\dot{\gamma} = \left(k\dot{\gamma}^{n-1}\right)\dot{\gamma} = k\dot{\gamma}^n$$

The graphs below show the shear stress and viscosity profiles using a Carreau model.

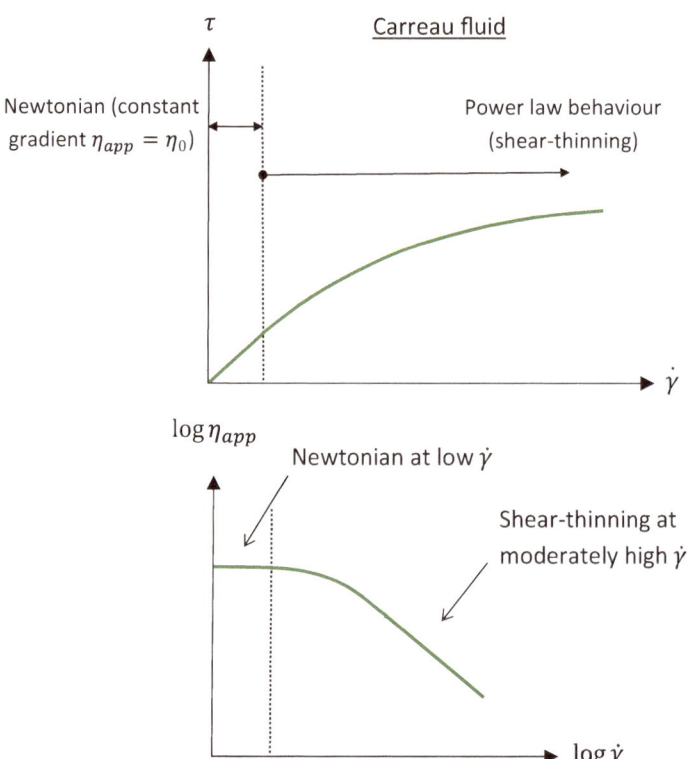

Another useful model to describe non-Newtonian fluids is the Cross model which has the form as shown below. In the equation below, α and m are constant parametric values.

$$\eta_{\text{app}} = \eta_\infty + \frac{\eta_0 - \eta_\infty}{1 + (\alpha\dot{\gamma})^m}$$

$$\tau = \eta_{\text{app}}\dot{\gamma} = \eta_\infty\dot{\gamma} + \left(\frac{\eta_0 - \eta_\infty}{1 + (\alpha\dot{\gamma})^m}\right)\dot{\gamma}$$

This model is different from the Carreau model in that it has an additional term, η_∞, which is viscosity at very high shear rates (also called the "constant infinite-shear-rate viscosity"), and it helps the model equation fit data in the Lower Newtonian region where viscosity "plateaus" to a constant value at very high shear rates.

The Cross model may therefore be used to describe fluids that exhibit three distinct behaviors:

1. Newtonian region at lower shear rates (Upper Newtonian region)
2. Power law region at intermediate shear rates
3. Newtonian region at very high shear rates (Lower Newtonian region)

The Cross model may be used to characterize fluids such as colloidal suspensions, polymer solutions, blood, or suspensions of carbon nanotubes in epoxy. In most cases, η_∞ is almost equivalent to (only slightly higher than) that of the base fluid (i.e., solvent or matrix) viscosity. In the context of a suspension, η_∞ may be understood as the viscosity of the base solvent of the suspension, and this base solvent is assumed Newtonian. Non-Newtonian behavior arises from interactions between particles suspended in the base fluid, and this interaction alters viscosity differently at different shear rates. The total viscosity η_{app} can be expressed as a sum of contributions from the interparticle interactions and the base fluid viscosity.

$$\eta_{app} = \eta_{base\ fluid} + \eta_{interactions} = \eta_\infty + \frac{\eta_0 - \eta_\infty}{1 + (\alpha\dot{\gamma})^m}$$

Upon inspection of the Cross equation, we observe three regions:

1. Constant viscosity η_0 at very low shear rates
2. Power law region at intermediate shear rates
3. Constant viscosity η_∞ at very high shear rates

Upper Newtonian Region

At low shear rates, we have Newtonian behavior where viscosity is a constant value.

$$(\alpha\dot{\gamma})^m \ll 1$$

$$\eta_{app} \cong \eta_\infty + \frac{\eta_0 - \eta_\infty}{[1]^m} = \eta_0$$

$$\tau = \eta_{app}\dot{\gamma} = \eta_0\dot{\gamma}$$

Power Law Region

At intermediate (moderately high) shear rates, we notice that the Cross equation simplifies to the power law model. Let us define $n - 1 = -m$, where n is a new parameter in the exponent term and let k be a constant such that $k = (\eta_0 - \eta_\infty)\alpha^{n-1}$.

$$(\alpha\dot{\gamma})^m \gg 1$$

$$\eta_{app} \cong \eta_\infty + \frac{\eta_0 - \eta_\infty}{(\alpha\dot{\gamma})^m} = \eta_\infty + \left[(\eta_0 - \eta_\infty)\alpha^{n-1}\right]\dot{\gamma}^{n-1} = \eta_\infty + k\dot{\gamma}^{n-1}$$

$$\eta_{app} - \eta_\infty = \eta_{interactions} = k\dot{\gamma}^{n-1}$$

In other words, with increasing shear rate in this region, the variation of incremental viscosity contributed by interparticle interactions (relative to the base fluid viscosity η_∞) follows a power law relationship.

Lower Newtonian Region

Finally, at very high shear rates, we have a total viscosity that plateaus to a constant value close to the viscosity of the base fluid of the suspension which we assume Newtonian. Therefore, viscosity can also be approximated as a constant value equivalent to η_∞.

$$\frac{\eta_0 - \eta_\infty}{1 + (\alpha\dot{\gamma})^m} \to 0$$

$$\eta_{app} \cong \eta_\infty = \eta_{\text{base fluid}}$$

The following plots show the shear stress and viscosity profiles for a Cross suspension.

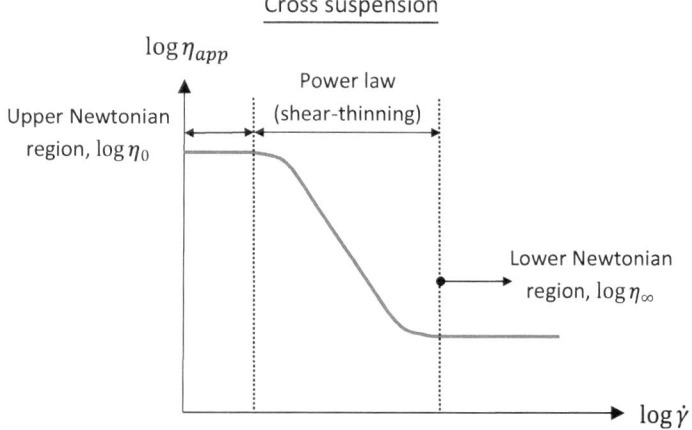

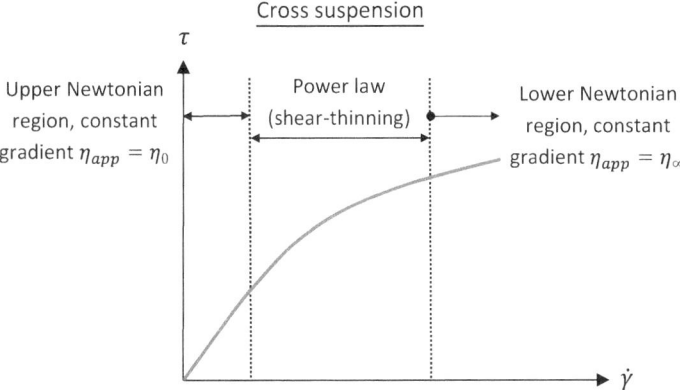

The three regions may also be understood at the molecular level. Most suspension-like systems contain particles with a certain size distribution. These particles may exist singly or in loose clusters, and examples of such particles include droplets in emulsions, bubbles in foams, and branched or entangled chains of polymer molecules in polymer solutions or melts.

Upper Newtonian Region

From a state of rest to very low shear rates, the particles orient randomly in the base fluid and take on an arrangement of the lowest energy state. For polymeric systems, polymer chains are highly entangled and coiled in a random orientation. At low shearing, the system resists deformation by offering high resistance which gives rise to a high viscosity or yield stress. This region is also called the Upper Newtonian region, whereby Newtonian behavior is observed with shear stress increasing linearly with an increasing shear (or strain) rate. This linear gradient also means a constant viscosity ("constant zero-shear-rate viscosity") denoted here as η_0.

Power Law Region

As shearing increases, the particles start to align themselves in the direction of shear (or flow). The fluid is said to become "pseudoplastic" as the particles suspended in the solvent deform as they orient along the streamlines. Clusters of particles may break up into smaller clusters or individual particles. As for polymeric systems, previously entangled chains of polymer molecules begin to disentangle and straighten out. These changes in particle arrangement facilitate flow, thus leading to a reduced apparent viscosity with increasing shear rate, i.e., shear-thinning. The Cross model simplifies to the power law model for this region.

Lower Newtonian Region

As shear rate increases further to very high values, the power law model and other similarly simplistic models may not be valid, and more rigorous models may need to be considered. For certain types of fluids such as polymer solutions, they depart from the shear-thinning region and enter the Lower Newtonian region, where flow behavior is Newtonian again with a constant observed viscosity η_∞. High shearing causes polymer chains to fully straighten out, the suspension becomes more homogeneously structured, and its viscosity gradually becomes more insensitive to changes in shear rate. The Lower Newtonian region is usually observed for polymer solutions. However, it is less common in polymer melts (especially thermoplastics), foams, emulsions, and suspensions as the very high shear rates cause the suspended molecules to degrade or polymer chains to break.

Problem 14

In rheometric experiments, fluid deformation caused by an applied constant shear or step shear leads to stress responses that are relatively simple to model. However in most cases, the shear (or strain) profile is non-linear with time and instead varies in an oscillatory manner.

(a) **Assuming a linear viscoelastic fluid and using a bimodal Maxwell model (comprising two Maxwell elements in parallel), derive the following expressions for the storage modulus G' and loss modulus G''. Explain what the symbols denote and the physical significance of the moduli.**

$$G' = \sum_{i=1}^{2} \frac{g_i \lambda_i^2 \omega^2}{1 + \lambda_i^2 \omega^2}, \quad G'' = \sum_{i=1}^{2} \frac{g_i \lambda_i \omega}{1 + \lambda_i^2 \omega^2}$$

(b) **The bimodal Maxwell model in part (a) was used to characterize a particular polymer melt whereby the relaxation times for the two modes are 2s and 0.004s, respectively. It was found that the storage modulus G' at the high-frequency limit is 110Pa and the complex viscosity η^* (expression given below) at the low-frequency limit is 1.8Pas.**

$$\eta^* = \frac{\sqrt{G'^2 + G''^2}}{\omega}$$

   (i) **Explain the physical meaning of the high- and low-frequency limits in describing the rheology of complex fluids.**
   (ii) **Determine the values of elastic constant g_i and viscosity η_i for the individual modes of the bimodal model.**

(iii) **Determine the angular frequency at which the complex viscosities of each mode are equivalent.**

(iv) **In a single log-log plot, sketch profiles for the complex viscosities of each mode, as well as that for the total bimodal system, against frequency. Comment on the shape of the plots.**

(v) **In a single log-log plot, sketch the expected profiles for complex viscosity η^* and storage and loss moduli G' and G'' for a typical polydisperse polymer melt. Comment on the plots with reference to the linear Newtonian region, the crossover point, and the shear-thinning region.**

Solution 14

Worked Solution

(a) First, let us represent our linear viscoelastic fluid using the Maxwell model. In a single mode Maxwell model, we mimic the viscous property of the fluid using a dashpot with viscosity η and the elastic property using a spring with elastic constant denoted g here. Both elements are connected in series as shown below for a single mode.

Spring   Dashpot
(elastic constant g)   (viscosity η)

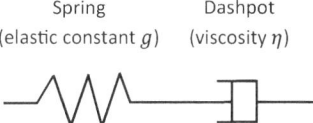

The expression for shear stress in the spring component follows Hooke's law.

$$\tau_{\text{spring}} = g\gamma_{\text{spring}}$$

The expression for shear stress in the dashpot component follows that for a perfectly Newtonian fluid.

$$\tau_{\text{dashpot}} = \eta\dot{\gamma}_{\text{dashpot}}$$

The overall shear stress τ for this Maxwell element is conceptually analogous to electric current in a circuit, while the overall strain γ is analogous to voltage. Therefore, the following equations apply.

$$\tau = \tau_{\text{spring}} = \tau_{\text{dashpot}} \tag{1}$$

$$\gamma = \gamma_{\text{spring}} + \gamma_{\text{dashpot}} \tag{2}$$

From Eq. 2 it is straightforward to deduce the following equation.

$$\dot{\gamma} = \dot{\gamma}_{\text{spring}} + \dot{\gamma}_{\text{dashpot}}$$

We can substitute the earlier expressions for strain rates and re-express the above in the form of a differential equation with respect to time.

$$\frac{d\gamma}{dt} = \frac{d}{dt}\left(\frac{\tau}{g}\right) + \frac{\tau}{\eta}$$

$$g\frac{d\gamma}{dt} = \frac{d\tau}{dt} + \left(\frac{g}{\eta}\right)\tau$$

Relaxation time λ is defined as follows and we can substitute into our differential equation above.

$$\lambda = \frac{\eta}{g}$$

$$g\frac{d\gamma}{dt} = \frac{d\tau}{dt} + \frac{\tau}{\lambda}$$

Now consider that we have an oscillatory shear rate which can be represented in terms of a strain profile written as a complex exponential function with angular frequency ω. γ_0 represents initial value of strain at time $t = 0$.

$$\gamma = \gamma_0 e^{i\omega t}$$

The above equation in fact originates from Euler's formula which is ubiquitously applied in science and engineering. It is useful as it relates oscillatory functions (e.g., trigonometric functions) to the complex exponential. In its general form, Euler's formula is expressed as follows where x and i are real and imaginary numbers, respectively, and x represents the argument of the trigonometric functions in radians.

$$e^{ix} = \cos x + i\sin x$$

In the case of an oscillatory shear rate (and hence oscillatory strain), it is easier to use complex numbers. Let us express our equations in the complex form, by considering stress response τ in terms of a complex modulus G^* that relates to shear stress in the same way that elastic constant g relates to τ in the Hooke's law equation $\tau = g\gamma$. Since G^* is a complex number, it can be expressed as the sum of a real part and an imaginary part.

$$\tau = G^* \gamma = (G' + iG'')\gamma$$

$$\tau = (G' + iG'')\gamma_0 e^{i\omega t}$$

We can substitute the above expression for stress into our earlier differential equation for a single mode Maxwell element. By definition, note that $i^2 = -1$.

$$g\frac{d\gamma}{dt} = \frac{d\tau}{dt} + \frac{\tau}{\lambda}$$

$$g\frac{d[\gamma_0 e^{i\omega t}]}{dt} = \frac{d[(G' + iG'')\gamma_0 e^{i\omega t}]}{dt} + \frac{(G' + iG'')\gamma_0 e^{i\omega t}}{\lambda}$$

$$g\gamma_0 i\omega e^{i\omega t} = (G' + iG'')\gamma_0 i\omega e^{i\omega t} + \frac{(G' + iG'')\gamma_0 e^{i\omega t}}{\lambda}$$

$$gi\omega = (G' + iG'')i\omega + \frac{G' + iG''}{\lambda} = (G'i\omega + i^2 G''\omega) + \frac{G' + iG''}{\lambda}$$

$$(g\omega)i = \left[G'\omega + \frac{G''}{\lambda} \right]i + \left(-G''\omega + \frac{G'}{\lambda} \right)$$

We can now equate the real part and imaginary part to obtain two separate equations 3 and 4. Note that on the left-hand side of the equation above, there is only an imaginary part, and the real part is therefore equivalent to zero.

$$-G''\omega + \frac{G'}{\lambda} = 0 \tag{3}$$

$$g\omega = G'\omega + \frac{G''}{\lambda} \tag{4}$$

From Eq. 3, we have the following expression for G' which we substitute into Eq. 4:

$$G' = \lambda\omega G''$$

$$g\omega = (\lambda G''\omega)\omega + \frac{G''}{\lambda}$$

$$\lambda g\omega = \lambda^2 G''\omega^2 + G''$$

$$G'' = \frac{g\lambda\omega}{1 + \lambda^2 \omega^2}$$

$$G' = \lambda\omega \left(\frac{g\lambda\omega}{1 + \lambda^2 \omega^2} \right) = \frac{g\lambda^2 \omega^2}{1 + \lambda^2 \omega^2}$$

Finally we note that in our problem we have a bimodal Maxwell model. In this setup, we connect two Maxwell elements in parallel. Shear stress τ is analgous to current in an electric circuit, which means that the total shear stress is the sum of shear stresses in the branches 1 and 2.

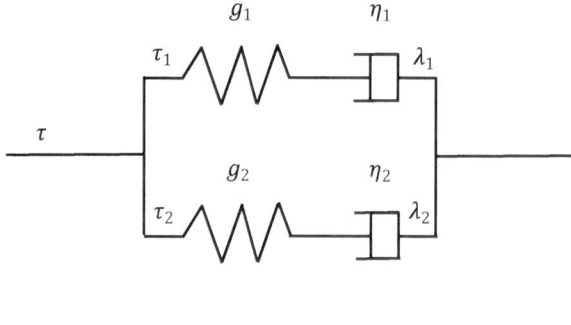

$$\tau = \tau_1 + \tau_2$$

We can now simply adjust our earlier expressions for storage modulus G' and loss modulus G'' by including a summation over two Maxwell elements to consider the two modes.

$$G' = \frac{g\lambda^2\omega^2}{1 + \lambda^2\omega^2} \rightarrow G' = \sum_{i=1}^{2} \frac{g_i\lambda_i^2\omega^2}{1 + \lambda_i^2\omega^2}$$

$$G'' = \frac{g\lambda\omega}{1 + \lambda^2\omega^2} \rightarrow G'' = \sum_{i=1}^{2} \frac{g_i\lambda_i\omega}{1 + \lambda_i^2\omega^2}$$

(b) (i) It is usually preferred to use a wide range of frequencies to study the characteristic flow patterns of complex fluids.

Low-Frequency Limit

At low frequencies, we approach a terminal flow limit. For fluids such as polymeric samples containing networks of long cross-linked chains, the low-frequency modulus increases with a higher density of cross-links and chain entanglements. At the low-frequency limit, complex viscosity data typically fits well for fluids such as water-oil emulsions, viscoelastic suspensions, and polymer solutions. However, this may not be the case for fluids such as micellar solutions that exhibit nonuniform flow or structural changes induced by flow. For multimodal systems, as angular frequency $\omega \rightarrow 0$, the fluid's complex viscosity η^* tends toward that of the individual mode with the most dominant relaxation time (highest value).

The low-frequency limit can be experimentally determined using methods such as the creep test, whereby a steady shear is applied. The resultant strain response and apparent viscosity can then be experimentally determined. Experimental methods in this case are however often limited by the long-term stability of the fluid as creep tests are more suitable for studying material response at longer timescales (minutes to days). For shorter timescales of a few seconds or less, it is more accurate to measure the material's dynamic response using a sinusoidal (or oscillatory) input load of strain or stress.

High-Frequency Limit

As frequencies increase, we enter the glassy regime and eventually reach a high-frequency limit also called the "rubber plateau." As angular frequency $\omega \to \infty$, the viscosity of the fluid tends to zero.

At higher frequencies, one is able to study a fluid's dynamic behavior, in terms of microstructure and molecular interactions such as in liquid colloidal suspensions or dispersions. For such suspensions, the applied shear force at high frequencies changes much faster than the characteristic relaxation time of the Brownian particles suspended in the fluid, and the storage modulus G' can be probed to study this high-frequency limit. For polymeric samples, the value of the storage modulus at high frequencies depends mainly on chain entanglements. Experiments can be conducted to determine the limiting storage modulus although the main challenge in this case lies in equipment limitations posed by rheometers.

(b) (ii) We obtained the following expression for storage modulus earlier.

$$G' = \sum_{i=1}^{2} \frac{g_i \lambda_i^2 \omega^2}{1 + \lambda_i^2 \omega^2} = \sum_{i=1}^{2} \frac{g_i \lambda_i^2}{\frac{1}{\omega^2} + \lambda_i^2}$$

At the high-frequency limit, $\omega \to \infty$ and $\frac{1}{\omega^2} \to 0$, therefore the expression for the storage modulus simplifies as follows:

$$G'\big|_{\omega \to \infty} = \sum_{i=1}^{2} g_i = g_1 + g_2$$

We are given the value of storage modulus at the high-frequency limit $G'\big|_{\omega \to \infty} = 110 Pa$, which we can substitute into the equation above.

$$110 = g_1 + g_2 \tag{1}$$

At the low-frequency limit, $\lambda_i^2 \omega^2 \ll 1$, therefore the expression for storage modulus simplifies as follows:

$$G' = \sum_{i=1}^{2} \frac{g_i \lambda_i^2 \omega^2}{1 + \lambda_i^2 \omega^2} \rightarrow G'|_{\omega \to 0} = \sum_{i=1}^{2} g_i \lambda_i^2 \omega^2$$

Similar for the loss modulus G'' at the low-frequency limit, $\lambda_i^2 \omega^2 \ll 1$; therefore the expression for loss modulus simplifies as follows:

$$G'' = \sum_{i=1}^{2} \frac{g_i \lambda_i \omega}{1 + \lambda_i^2 \omega^2} \rightarrow G''|_{\omega \to 0} = \sum_{i=1}^{2} g_i \lambda_i \omega$$

Given the following expression for complex viscosity and its value of 1.8Pas at the low-frequency limit, we have the following:

$$\eta^*|_{\omega \to 0} = \frac{\sqrt{G'^2 + G''^2}}{\omega} = 1.8$$

$$\sqrt{\left(\sum_{i=1}^{2} g_i \lambda_i^2 \omega^2\right)^2 + \left(\sum_{i=1}^{2} g_i \lambda_i \omega\right)^2} = 1.8\omega$$

$$\sqrt{\left(g_1 \lambda_1^2 \omega^2 + g_2 \lambda_2^2 \omega^2\right)^2 + \left(g_1 \lambda_1 \omega + g_2 \lambda_2 \omega\right)^2} = 1.8\omega$$

$$\sqrt{\left[\omega^2 \left(g_1 \lambda_1^2 + g_2 \lambda_2^2\right)\right]^2 + \left[\omega(g_1 \lambda_1 + g_2 \lambda_2)\right]^2} = 1.8\omega$$

Since $\omega \to 0$, the term $(\omega^2(g_1\lambda_1^2 + g_2\lambda_2^2))^2 \ll (\omega(g_1\lambda_1 + g_2\lambda_2))^2$ since the diminishing effect of ω^2 outweighs that of ω in the terms. Therefore we can further simplify our expression to the following:

$$\sqrt{\left[\omega(g_1 \lambda_1 + g_2 \lambda_2)\right]^2} = 1.8\omega$$

$$g_1 \lambda_1 + g_2 \lambda_2 = 1.8$$

We are given the relaxation times for the two modes in the bimodal model, $\lambda_1 = 2s$ and $\lambda_2 = 0.004s$ which we can now substitute into the expression above.

$$2g_1 + 0.004g_2 = 1.8 \tag{2}$$

We now have two simultaneous Eqs. 1 and 2 which we can use to solve for the two unknowns g_1 and g_2.

$$2(110 - g_2) + 0.004g_2 = 1.8$$

$$220 - 1.8 = 1.996g_2 \rightarrow g_2 = 109Pa$$

$$g_1 = 110 - g_2 = 0.68 Pa$$

We can now determine the viscosities of each mode by definition.

$$\eta_1{}^* = g_1 \lambda_1 = 0.68(2) = 1.4 Pas$$

$$\eta_2{}^* = g_2 \lambda_2 = 109(0.004) = 0.44 Pas$$

(b) (iii) The condition whereby the complex viscosities of each mode are equal is expressed below.

$$\eta_1{}^* = \eta_2{}^*$$

$$\frac{\sqrt{G_1'{}^2 + G_1''{}^2}}{\omega} = \frac{\sqrt{G_2'{}^2 + G_2''{}^2}}{\omega}$$

$$\sqrt{G_1'{}^2 + G_1''{}^2} = \sqrt{G_2'{}^2 + G_2''{}^2}$$

$$\sqrt{\left(\frac{g_1 \lambda_1{}^2 \omega^2}{1 + \lambda_1{}^2 \omega^2}\right)^2 + \left(\frac{g_1 \lambda_1 \omega}{1 + \lambda_1{}^2 \omega^2}\right)^2} = \sqrt{\left(\frac{g_2 \lambda_2{}^2 \omega^2}{1 + \lambda_2{}^2 \omega^2}\right)^2 + \left(\frac{g_2 \lambda_2 \omega}{1 + \lambda_2{}^2 \omega^2}\right)^2}$$

$$\sqrt{\frac{g_1{}^2 \lambda_1{}^4 \omega^4 + g_1{}^2 \lambda_1{}^2 \omega^2}{\left(1 + \lambda_1{}^2 \omega^2\right)^2}} = \sqrt{\frac{g_2{}^2 \lambda_2{}^4 \omega^4 + g_2{}^2 \lambda_2{}^2 \omega^2}{\left(1 + \lambda_2{}^2 \omega^2\right)^2}}$$

$$\sqrt{\frac{g_1{}^2 \lambda_1{}^2 \omega^2 \left(\lambda_1{}^2 \omega^2 + 1\right)}{\left(1 + \lambda_1{}^2 \omega^2\right)^2}} = \sqrt{\frac{g_2{}^2 \lambda_2{}^2 \omega^2 \left(\lambda_2{}^2 \omega^2 + 1\right)}{\left(1 + \lambda_2{}^2 \omega^2\right)^2}}$$

$$\frac{g_1 \lambda_1}{\sqrt{1 + \lambda_1{}^2 \omega^2}} = \frac{g_2 \lambda_2}{\sqrt{1 + \lambda_2{}^2 \omega^2}}$$

Substituting the values we found earlier from part (b) (ii), we find the angular frequency at which complex viscosities of both modes are equal.

$$\frac{0.68(2)}{\sqrt{1 + 2^2 \omega^2}} = \frac{109(0.004)}{\sqrt{1 + (0.004)^2 \omega^2}} \rightarrow \omega = 1.48 rad/s$$

(b) (iv)

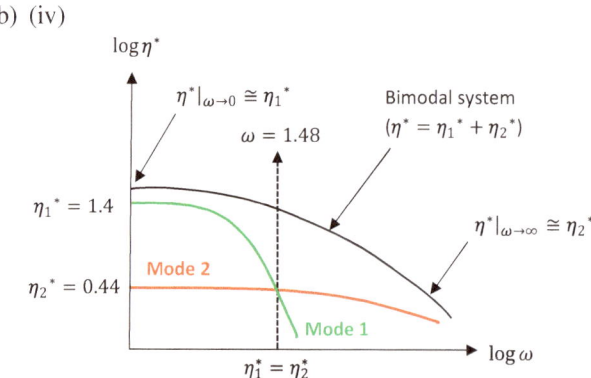

We note that for our bimodal system, the total complex viscosity of the fluid is the sum of its component modes.

$$\eta^* = \eta_1^* + \eta_2^*$$

Bimodal systems may not be rigorous enough to model real polymeric systems, especially more complex forms such as those involving polydisperse polymers which are commonly found in practical applications. In this example, only two modes are used. A better fit to real data for linear viscoelastic polymers is expected if more nodes are included in the model. For multiple nodes, the relaxation times of each node λ_i will need to be determined.

(b) (v)

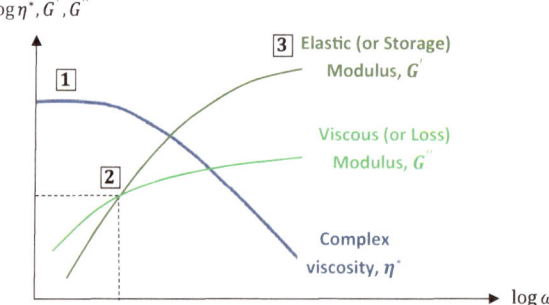

This is a typical plot describing the rheological behavior of a polydisperse polymer melt in a "frequency sweep" experiment. In such experiments, the angular frequency is varied at a fixed strain rate (or shear rate). The magnitude of strain applied in such experiments is usually well controlled at a small enough value such that the viscoelastic fluid is maintained in the linear Newtonian region. This prevents large strains from deforming the fluid, causing structural changes which may complicate experimental results. [Note: Another type of experiment that is commonly done is the "strain sweep" which involves a varying strain rate at a constant frequency.]

There are a few features to note about this plot, and they are enumerated in the annotations in the plot above, with explanations below.

1. At low frequencies, we observe a horizontal plateau for complex viscosity that fits Newtonian behavior where there is constant viscosity with increasing angular frequency within this range. At low frequencies, we also observe that the loss modulus is greater than the storage modulus, $G'' > G'$ which means that the fluid is largely viscous, with viscous dissipation as heat energy.

2. The crossover point is where $G' = G''$, and it represents the transition point from an elastic regime (dominates at high frequencies) to the terminal flow or Newtonian regime (dominates at low frequency). The frequency at which this crossover occurs is also called the "crossover frequency."

3. At higher frequencies, we observe that complex viscosity starts to decrease with increasing frequency. This phenomenon is also known as shear-thinning and can be fit to models such as the power law and Carreau models. In this region, viscosity effects are outweighed by elastic effects, and it makes sense therefore that the storage modulus dominates over the loss modulus whereby $G' > G''$. The fluid is said to be dominantly elastic, exhibiting non-linearity at higher frequencies. [Note: In this case we used the example of a shear-thinning fluid. However it is possible for a viscoelastic shear-thickening fluid to similarly exhibit a dominantly viscous region at low frequencies and a dominantly elastic region at higher frequencies in a frequency sweep experiment. One example is "Silly Putty," a silicone-based polymer suspension which consists of a large network of cross-linked chains.]

Coordinate Frames and the Stress Matrix

Problem 15

A stress matrix is commonly seen in the study of rheology. Consider a polymer flow with the stress matrix (in kPa) defined in coordinate frame 1 as shown below.

Coordinate Frame 1

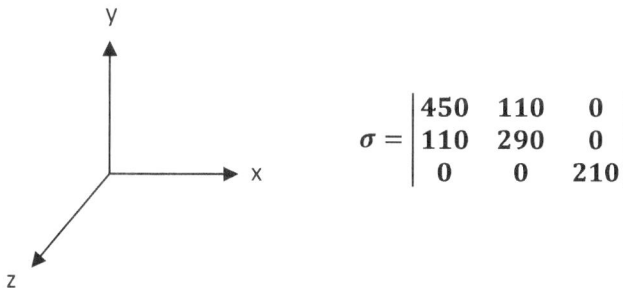

$$\sigma = \begin{vmatrix} 450 & 110 & 0 \\ 110 & 290 & 0 \\ 0 & 0 & 210 \end{vmatrix}$$

(a) Using suitable diagrams, explain what the values in the stress matrix mean in terms of forces acting on a fluid element.

(b) Determine the following for a new coordinate frame 2 whereby the x-axis is rotated 15° anticlockwise in the x-y plane (assuming the axis convention as illustrated above).

    (i) Direction cosine matrix between coordinate frames 1 and 2.

© Springer Nature Switzerland AG 2019
X. W. Ng, *Pocket Guide to Rheology: A Concise Overview and Test Prep for Engineering Students*, https://doi.org/10.1007/978-3-030-30585-7_3

(ii) **Full stress matrix in the new coordinate frame 2.**
(iii) **First and second invariants for the coordinate frames 1 and 2. Comment on your result.**

Solution 15

Worked Solution

(a) Let us visualize the physical meaning of values in the stress matrix. Imagine a fluid element as illustrated below in the old coordinate frame (i.e., frame 1), the stress matrix then provides information about the stress forces acting on the element. Using the x-face as an example, whereby the x-face is defined as the plane that has a normal vector pointing in the x-direction, the forces acting on the x-face are therefore 110 kPa along the x-plane (i.e., in the positive y-direction) and 450 kPa outward from the x-plane (in the positive x-direction). These forces are indicated in green below.

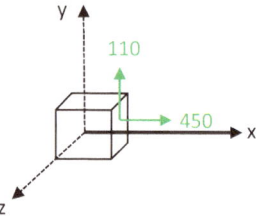

For the y-face, which is defined as the plane that has a normal vector pointing in the y-direction, it follows in the same way that the forces acting on this face are 110 kPa along the y-plane (in the positive x-direction) and 290 kPa outward from the y-plane (in the positive y-direction).

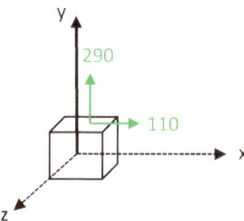

Finally, for the z-face, defined as the plane that has a normal vector pointing in the z-direction, there is only one force of magnitude 210 kPa acting outward from the z-plane (in the positive z-direction).

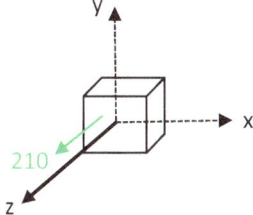

(b) (i) We now have a new coordinate frame 2, which is rotated 15° anticlockwise from frame 1 as shown below. Note that since the rotation is done in the x-y plane, the plane essentially rotates about the z-axis, and therefore the new z-axis (denoted "z' axis" here) points in the same direction as the z-axis in the old frame 1.

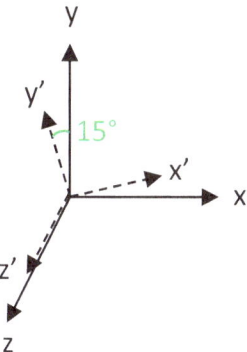

The direction cosine matrix can be expressed as follows:

$$\begin{vmatrix} \cos 15^\circ & \cos\left(90^\circ - 15^\circ\right) & 0 \\ \cos\left(90^\circ + 15^\circ\right) & \cos 15^\circ & 0 \\ 0 & 0 & \cos 0^\circ \end{vmatrix} = \begin{vmatrix} 0.966 & 0.259 & 0 \\ -0.259 & 0.966 & 0 \\ 0 & 0 & 1 \end{vmatrix}$$

(b) (ii) Before we proceed further, let us take note of certain key expressions and conventions for the stress matrix (or stress tensor). In general, the stress matrix in a new coordinate frame is denoted σ'_{ij} and expressed as follows, whereby a_{ik} and a_{jl} are the directional cosines, and σ_{kl} is the stress tensor in the old coordinate frame.

$$\sigma'_{ij} = \begin{vmatrix} \sigma'_{xx} & \sigma'_{xy} & \sigma'_{xz} \\ \sigma'_{yx} & \sigma'_{yy} & \sigma'_{yz} \\ \sigma'_{zx} & \sigma'_{zy} & \sigma'_{zz} \end{vmatrix} = a_{ik} a_{jl} \sigma_{kl} = \cos\theta_{ik} \cos\theta_{jl} \sigma_{kl}$$

The directional cosines store information about the rotation between the two frames. Note that by convention, the first and second subscripts of the terms a_{ik} and $\cos\theta_{ik}$ refer to the new and old frames, respectively. For example, in this problem, θ_{xy} denotes the angle between the x-axis in the new coordinate frame 2 and the y-axis in old coordinate frame 1. Note also that for the stress matrix, denoted σ_{xy} in this problem, the first subscript refers to the face (normal vector to this face is the x-axis), and the second subscript refers to the direction of the force acting on that face (pointing in the positive y-direction).

In our problem, stress matrix in coordinate frame 1, σ takes on the following values.

$$\sigma = \begin{vmatrix} \sigma_{xx} & \sigma_{xy} & \sigma_{xz} \\ \sigma_{yx} & \sigma_{yy} & \sigma_{yz} \\ \sigma_{zx} & \sigma_{zy} & \sigma_{zz} \end{vmatrix} = \begin{vmatrix} 450 & 110 & 0 \\ 110 & 290 & 0 \\ 0 & 0 & 210 \end{vmatrix}$$

The new stress matrix in coordinate frame 2 is denoted σ' here and expressed as shown below.

$$\sigma' = \begin{vmatrix} \sigma'_{xx} & \sigma'_{xy} & \sigma'_{xz} \\ \sigma'_{yx} & \sigma'_{yy} & \sigma'_{yz} \\ \sigma'_{zx} & \sigma'_{zy} & \sigma'_{zz} \end{vmatrix}$$

From the direction cosines computed in part i, we can also express the matrix form of the directional cosines as follows.

$$\begin{vmatrix} a_{xx} & a_{xy} & a_{xz} \\ a_{yx} & a_{yy} & a_{yz} \\ a_{zx} & a_{zy} & a_{zz} \end{vmatrix} = \begin{vmatrix} 0.966 & 0.259 & 0 \\ -0.259 & 0.966 & 0 \\ 0 & 0 & 1 \end{vmatrix}$$

We notice that there are four zero values in the directional cosine matrix above, $a_{xz} = a_{yz} = a_{zx} = a_{zy} = 0$; it follows therefore that $\sigma'_{xz} = \sigma'_{yz} = \sigma'_{zx} = \sigma'_{zy} = 0$ since their coefficients (i.e., the directional cosine values) are zero. We can now simplify our stress matrix in coordinate frame 2 as follows.

$$\sigma' = \begin{vmatrix} \sigma'_{xx} & \sigma'_{xy} & 0 \\ \sigma'_{yx} & \sigma'_{yy} & 0 \\ 0 & 0 & \sigma'_{zz} \end{vmatrix}$$

We have five remaining values σ'_{xx}, σ'_{xy}, σ'_{yx}, σ'_{yy}, and σ'_{zz} to be determined. These values will be computed from the corresponding directional cosines and values from the stress matrix in coordinate frame 1 (i.e., $\sigma_{xx}, \sigma_{xy}, \sigma_{yx}, \sigma_{yy}$ and σ_{zz}). (Note: color coding applied here subsequently is to help understand the notations used in the calculation steps below.)

We now observe from the values in our stress matrix σ that it is symmetric along the diagonal, $\sigma_{kl} = \sigma_{lk}$. It follows that the new stress matrix in coordinate frame 2 will also be symmetric and $\sigma'_{ij} = \sigma'_{ji}$. This allows us to "shortcut" the calculation process as we need only to compute four values instead of five, since $\sigma'_{xy} = \sigma'_{yx}$ for a symmetric matrix. We can now go about computing the remaining unknown values.

$$\sigma'_{xx} = a_{xk}a_{xl}\sigma_{kl} = a_{xx}a_{xx}\sigma_{xx} + a_{xx}a_{xy}\sigma_{xy} + a_{xy}a_{xx}\sigma_{yx} + a_{xy}a_{xy}\sigma_{yy} + a_{xz}a_{xz}\sigma_{zz}$$

$$\sigma'_{xx} = 0.966(0.966)(450) + 0.966(0.259)(110) + 0.259(0.966)(110) \\ + 0.259(0.259)(290) + 0(0)(210)$$

$$\sigma'_{xx} = 494$$

We can repeat the same process to find all other values in the stress matrix σ':

$$\sigma'_{xy} = a_{xx}a_{yx}\sigma_{xx} + a_{xx}a_{yy}\sigma_{xy} + a_{xy}a_{yx}\sigma_{yx} + a_{xy}a_{yy}\sigma_{yy} + a_{xz}a_{yz}\sigma_{zz}$$

$$\sigma'_{xy} = 0.966(-0.259)(450) + 0.966(0.966)(110) + 0.259(-0.259)(110) \\ + 0.259(0.966)(290) + 0(0)(210)$$

$$\sigma'_{xy} = 55$$

$$\sigma'_{yx} = \sigma'_{xy} = 55$$

$$\sigma'_{yy} = a_{yx}a_{yx}\sigma_{xx} + a_{yx}a_{yy}\sigma_{xy} + a_{yy}a_{yx}\sigma_{yx} + a_{yy}a_{yy}\sigma_{yy} + a_{yz}a_{yz}\sigma_{zz}$$

$$\sigma'_{yy} = -0.259(-0.259)(450) + (-0.259)(0.966)(110) + 0.966(-0.259)(110) \\ + 0.966(0.966)(290) + 0(0)(210)$$

$$\sigma'_{yy} = 246$$

$$\sigma'_{zz} = a_{zx}a_{zx}\sigma_{xx} + a_{zx}a_{zy}\sigma_{xy} + a_{zy}a_{zx}\sigma_{yx} + a_{zy}a_{zy}\sigma_{yy} + a_{zz}a_{zz}\sigma_{zz}$$

$$\sigma'_{zz} = 0(0)(450) + 0(0)(110) + 0(0)(110) + 0(0)(290) + 1(1)(210)$$

$$\sigma'_{zz} = 210$$

We can now substitute all values into the stress matrix for coordinate frame 2 to obtain the result as shown. Notice that the new stress matrix is symmetric.

$$\sigma' = \begin{vmatrix} \sigma'_{xx} & \sigma'_{xy} & 0 \\ \sigma'_{yx} & \sigma'_{yy} & 0 \\ 0 & 0 & \sigma'_{zz} \end{vmatrix} = \begin{vmatrix} 494 & 55 & 0 \\ 55 & 246 & 0 \\ 0 & 0 & 210 \end{vmatrix}$$

We can verify that our earlier assumption ($\sigma'_{yx} = \sigma'_{xy}$) was right with a simple extra calculation as shown below.

$$\sigma'_{yx} = a_{yx}a_{xx}\sigma_{xx} + a_{yx}a_{xy}\sigma_{xy} + a_{yy}a_{xx}\sigma_{yx} + a_{yy}a_{xy}\sigma_{yy} + a_{yz}a_{xz}\sigma_{zz}$$

$$\sigma'_{yx} = -0.259(0.966)(450) + (-0.259)(0.259)(110) + 0.966(0.966)(110)$$
$$+ 0.966(0.259)(290) + 0(0)(210)$$

$$\sigma'_{yx} = 55 = \sigma'_{xy}$$

(b) (iii) The first and second invariants, denoted here as I_1 and I_2, respectively, are termed invariants as they represent properties of the stress matrix that are independent of the choice of coordinate frame. They are expressed as follows for the matrix in coordinate frame 1.

$$I_{1,\text{frame }1} = \text{Trace } \sigma = \sigma_{xx} + \sigma_{yy} + \sigma_{zz}$$

$$I_{1,\text{frame }1} = 450 + 290 + 210 = 950$$

$$I_{2,\text{frame }1} = \sigma_{xx}\sigma_{yy} + \sigma_{yy}\sigma_{zz} + \sigma_{xx}\sigma_{zz} - \sigma_{xy}\sigma_{yx} - \sigma_{yz}\sigma_{zy} - \sigma_{xz}\sigma_{zx}$$

$$I_{2,\text{frame }1} = 450(290) + 290(210) + 450(210) - 110(110) - 0(0) - 0(0)$$
$$\cong 274000$$

And for coordinate frame 2,

$$I_{1,\text{frame }2} = \text{Trace } \sigma' = \sigma'_{xx} + \sigma'_{yy} + \sigma'_{zz}$$

$$I_{1,\text{frame }2} = 494 + 246 + 210 = 950$$

$$I_{2,\text{frame }2} = \sigma'_{xx}\sigma'_{yy} + \sigma'_{yy}\sigma'_{zz} + \sigma'_{xx}\sigma'_{zz} - \sigma'_{xy}\sigma'_{yx} - \sigma'_{yz}\sigma'_{zy} - \sigma'_{xz}\sigma'_{zx}$$

$$I_{2,\text{frame }2} = 494(246) + 246(210) + 494(210) - 55(55) - 0(0) - 0(0) \cong 274000$$

We have shown here that regardless of coordinate frame, the values of the first and second invariants are the same.

Problem 16

Explain the downsides of using scalar quantities to model real fluids in practical applications and suggest improvements to the scalar approach.

Solution 16

Worked Solution

Most fluids in real-life applications are complex fluids that do not display straight-forward flow behaviors. The fluids themselves may be morphologically complex, and therefore bear unique intrinsic properties when deformed or set in motion. Under different circumstances (e.g., different deformation rates), the same fluid may also exhibit different flow patterns. Such complex cases are often not sufficiently modeled using scalar approaches, and therefore it is necessary to consider multidimensional methods involving vectors and matrices.

When we say "complex fluids," we are often referring to non-Newtonian fluids. For recap, a Newtonian fluid exhibits the following shear stress τ correlation with shear rate (or strain rate) $\dot{\gamma}$.

$$\tau = \eta\dot{\gamma}$$

Newtonian fluids are simplest to understand and model as they exhibit linear behavior with stress scaling linearly with shear rate according to a proportionality constant that represents viscosity η. It is in this context that scalar constitutive equations are sufficiently suitable for modeling Newtonian fluids for simple flow conditions. Common examples of Newtonian fluids include water and mineral oil.

However, in most processing applications, we encounter non-Newtonian or complex fluids. Moreover, flow conditions may be complicated such that scalar equations no longer represent even Newtonian fluids well in such situations. One example that illustrates the importance of flow conditions is in how certain fluids behave differently under shear and extensional flows. A more generalized and rigorous method is necessary to better describe such flow processes completely.

Stress and strain rate are in their broader definitions, tensor quantities which may be denoted σ_{ij} and $\dot{\gamma}_{kl}$, respectively. It is common for stresses to be further classified as a normal or shear stress whereby the former acts perpendicular to the plane while the latter acts parallel to the plane. In this case, the notations used for normal and shear stresses would be σ_{ij} and τ_{ij}, respectively. Constitutive equations relate stress to strain or strain rate; therefore, it follows that quantities of strain and strain rate may also be better represented using tensors.

Tensor quantities, unlike scalars, are more than one-dimensional and are directional. This makes tensors more suitable for complex fluids and engineering flows as three-dimensional (3D) solutions can be obtained. When we describe stress on a fluid particle (assume a small cubic element), we may holistically describe its nine components in the form of a three-dimensional vector (e.g., x-y-z axes) or 3 by 3 matrix, which contains nine quantities, each with its unique direction and value. For an x-y-z coordinate system in Cartesian form, the values that subscripts of σ_{ij}, τ_{ij}, and $\dot{\gamma}_{ij}$ take will therefore also be x, y, and z.

Consider a cubic element of fluid particle, the stress matrix may be expressed as follows whereby the diagonal quantities represent normal stresses σ_{ii} that act perpendicular to the cube face, while the rest of the quantities represent shear stresses τ_{ij} acting parallel to the cube face. This tensor representation for stress is more complete than a scalar quantity, as it can hold information about stresses acting on all six faces of the cube and in all directions described by the 3D axes.

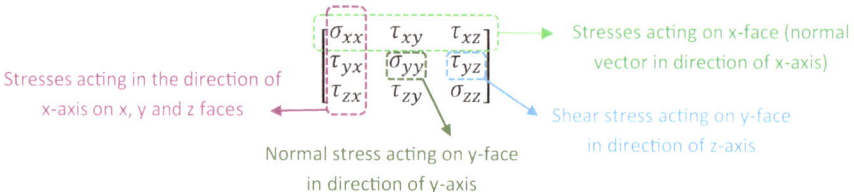

We can better visualize the physical significance of some of the matrix elements mentioned above by illustrating them as stress vectors as follows.

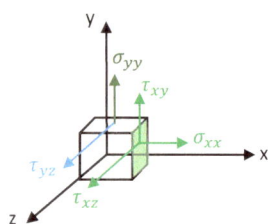

Problem 17

A dilute glucose solution was subject to shear flow in the $x - y$ plane and was observed to behave like a Newtonian fluid. It was found that the principal stress difference denoted $\sigma^*_{xx} - \sigma^*_{yy}$ occurs at a direction of principal stress components and at an angle of $\frac{\pi}{4}$ radians to the direction of flow in the horizontal x direction:

(a) Explain what is meant by "principal stress," using suitable diagrams.
(b) Mohr's circle presents a graphical representation of tensor transformations and is extremely useful in visualizing coordinate frame rotations. Given that the normal stress difference in the laboratory frame (i.e., $\sigma_{xx} - \sigma_{yy}$) is equivalent to zero, comment on the significance of this result using a Mohr's circle.
(c) Show that the normal stress difference in the lab frame is zero using the tensor transformation formula as shown below, where σ^{new}_{ij} and σ^{old}_{kl} denote stress tensors in the new and old coordinate frames, respectively, and a_{ik} and a_{jl} denote direction cosines describing the transformation between the two frames,

$$\sigma^{new}_{ij} = a_{ik}a_{jl}\sigma^{old}_{kl}$$

(d) Given that the magnitude of the principal stress difference is 180 Pa for a simple shear rate of 4 s^{-1}, find the viscosity of the fluid.
(e) A polymer is now added to the solution which increased the magnitude of the principal stress difference by 2.3 times, and the direction of the new principal stress is now 25° to the direction of flow:

  (i) Determine the magnitude of normal stress difference in the laboratory frame.
  (ii) Determine the shear stress in the lab frame and the apparent viscosity of this solution at a shear rate of 4 s^{-1}. Comment on the effects of polymer addition.

(f) Briefly comment on how an understanding of stress difference and tensor transformation techniques might be useful in studying the rheological behavior of fluids such as sugar solutions in practical applications.

Solution 17

Worked Solution

(a) For an arbitrary fluid or material in general, we can imagine taking an infinites-
imally small cubic element of it such that we have six faces or planes. Assuming
a two-dimensional system, the normal stresses denoted σ (acting perpendicular
to the planes) and the shear stresses denoted τ (acting parallel to the planes) may
be illustrated below left. It is worth noting that by sign convention, tensile
stresses which point out of the plane are positive, while compressive stresses
that point into the plane are negative. The first subscript refers to the normal
vector of the face at which the stress is acting, and the second subscript refers to
the direction of the stress.

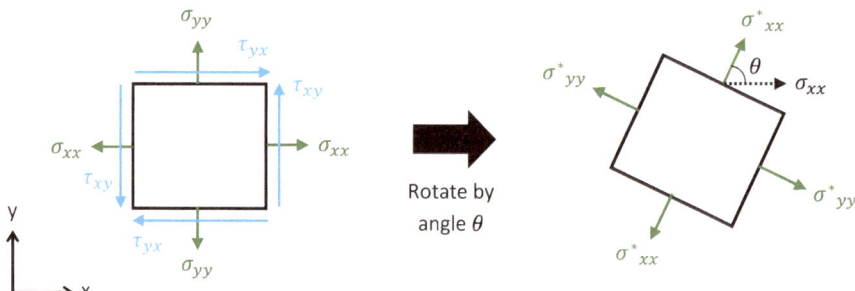

Principal stresses denoted σ^* are shown in the diagram above on the right, and they
refer to normal stresses acting perpendicular to pairs of mutually orthogonal
(or perpendicular) planes. We note particularly that we call the normal stresses
"principal stresses" when we are in the "principal frame" whereby there are no
shear stresses (acting parallel to the plane) present. We can arrive at the principal
frame by rotating the element about the z-axis by an angle θ as shown in the above
diagram, until the shear stresses are zero.

The non-zero normal stresses remaining after this rotation are "principal stresses,"
and the one with the maximum absolute value (ignoring signs) is known as the
"major principal stress," while the one with the minimum absolute value is called the
"minor principal stress." The difference between these two values is also known as
the "principal stress difference."

In practical applications, it is useful to know at which planes and what angles of
inclination that maximum and/or minimum stresses (be it normal or shear stress)
occur, as it helps us understand failure limits for materials and identify stability
regions of fluids.

(b) We can construct a Mohr's circle for our problem, but before that, let us first consider a general case (consisting of both tension and compression) to introduce key concepts of the Mohr's circle.

In a Mohr's circle, the vertical axis denotes shear stress τ, while the horizontal axis denotes normal stress σ. Let us consider stresses acting on two mutually orthogonal planes (e.g., x and y planes) of a small square element of fluid. Assuming that the stresses have values of $\sigma_{xx} = 4$ (positive in tension), $\sigma_{yy} = -2$ (negative in compression), and $|\tau_{xy}| = |\tau_{yx}| = 3$, we arrive at the Mohr's circle construction as shown below where the two points occur on the x and y faces, respectively.

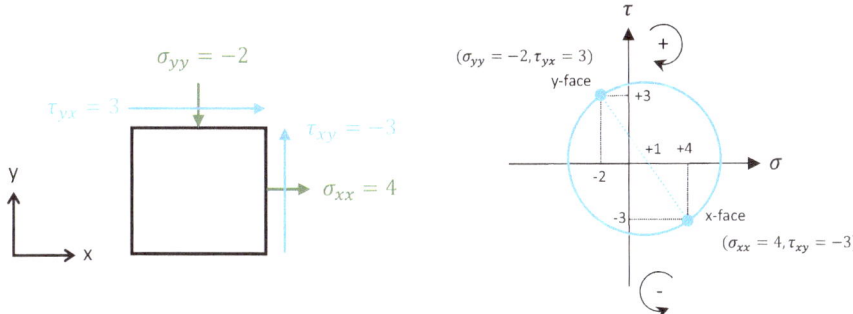

Note that for the Mohr's circle only, we define shear stress as positive when it acts to rotate the element clockwise and negative when it acts to rotate the element anticlockwise. Therefore, the stress acting on the y-face τ_{yx} is positive, with a direction pointing to the right and along the y-face as shown above left. The shear stress acting on the x-face is equal in magnitude as the shear stress on the y-face but is opposite in sign, i.e., it contributes to an anticlockwise rotation.

We note that the line connecting the two points represents the x and y faces which are mutually orthogonal and hence $90°$ apart. On the Mohr's circle, this angle will be reflected as twice the value of $90°$; therefore we see that $180°$ separates the two points representing the x and y planes on the circle. Essentially, the line connecting the two points is also the diameter of the circle, and it crosses the horizontal σ-axis at the midpoint value which can be computed as $\sigma_{mid} = 0.5(\sigma_{xx} + \sigma_{yy}) = 1$ which is indicated in the diagram above (right).

Let us now consider rotating the coordinate frame for this general example, by rotating the axes anticlockwise over an angle θ to a new coordinate system of x'-y'. The corresponding changes to the Mohr's circle construction can be seen below. What we now observe is that $\sigma_{x'x'}$ is still positive (i.e., in tension), $\sigma_{y'y'}$ is still negative (i.e., in compression), but $\tau_{x'y'}$ is now positive (i.e., acts to rotate the element clockwise). Therefore we have the corresponding force diagram as shown below on the right.

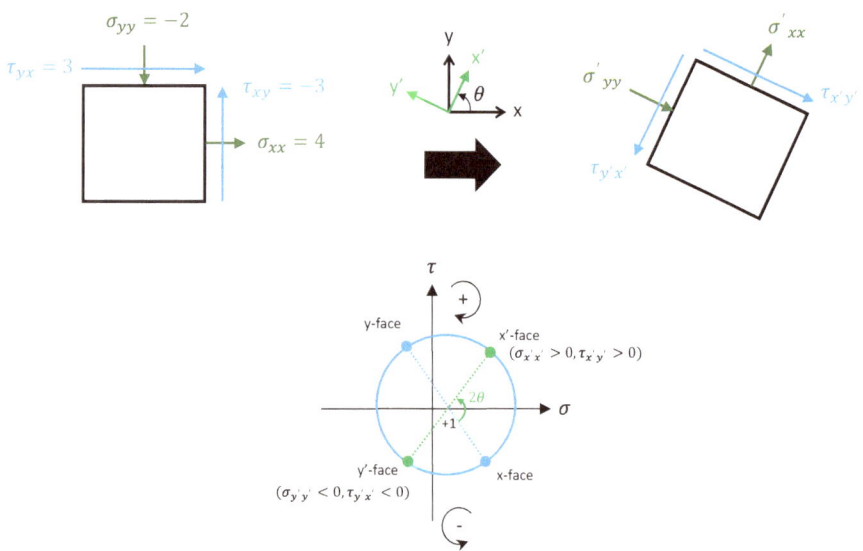

Special Cases

Now let us examine two special cases, the first case is shown below left whereby the rotation brings the two points (representing the x and y faces respectively) to fall exactly on the σ-axis, and the second case is shown below right when the two points fall exactly on the τ-axis. We note that when shear stresses are zero, the absolute values of the non-zero normal stresses are maximum. In this case, we are in the principal frame whereby we have maximum values for normal stresses, and the corresponding x and y planes are also the principal planes. On the contrary, for the diagram below right, we have the opposite case whereby the shear stresses are at their maximum magnitudes and the normal stresses take on the same value. The physical significance of these special cases can be further developed by considering the example in our problem.

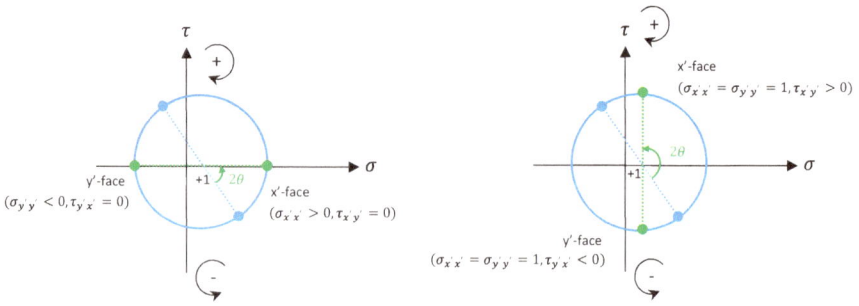

Let us now return to our problem, where we have a shear flow as shown below, observed in the lab frame.

We are told that the normal stress difference $\sigma_{xx} - \sigma_{yy} = 0$ in the lab frame, which means that the Mohr's circle will have a diameter that now passes through the origin as shown below. We note further that there is an angle of $\frac{\pi}{4}$ radians between the flow direction and the principal horizontal axis. We can deduce that an angle of $\frac{\pi}{4}$ radians translates into twice the value, which is $\frac{\pi}{2}$ radians between the principal frame (denoted with $*$) and the lab frame. Combining this knowledge with the fact earlier established that shear stresses are zero in the principal frame, we can construct the principal frame Mohr line as a horizontal line as shown in green and the lab frame Mohr line as shown in blue. With maximum shear stress occurring when we have the blue line, we arrive at the second special case as described earlier. The principal plane represented by the green line corresponds to the first special case as described earlier.

In this problem, another feature we observe is that normal stress difference is zero. This condition is also known as "pure shear" which is a special case of simple shear flow whereby there is no rotation (no bending moments).

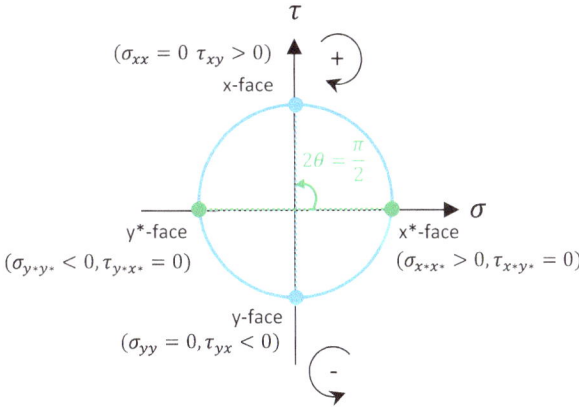

There are a few positions that Mohr's circles can take, and other than the case of pure shear when the center of the circle is at the origin. Some examples and what they mean are shown below. Note that the position of the circle provides information about the stresses involved, while the specific line connecting two diametrically opposite points on the circle's circumference may take any position as it rotates about the center of the circle according to the choice of coordinate frame we choose to analyze the flow condition.

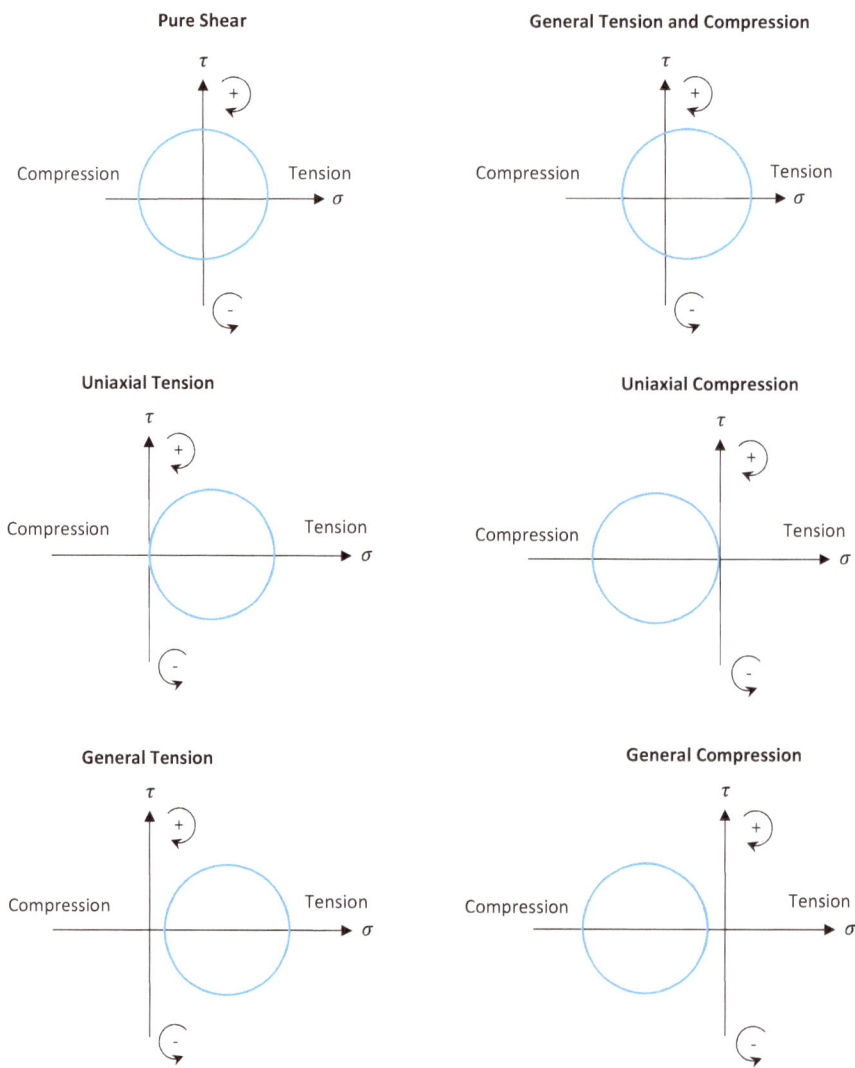

(c) We can mathematically show that the normal stress difference in the lab frame is zero using the tensor transformation formula given as shown below.

$$\sigma^{new}_{ij} = a_{ik} a_{jl} \sigma^{old}_{kl}$$

In the expression above, let us define the new frame as the lab frame (since it is the unknown, we need to compute/proof) and let the old frame be the principal frame (since we are given starting information in the problem about the principal frame having an angle of $\frac{\pi}{4}$ radians between the flow direction and the horizontal principal axis).

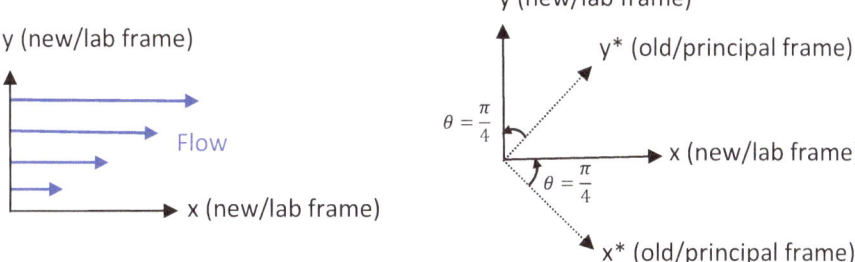

With the information above, we can construct the Mohr's circle as shown below.

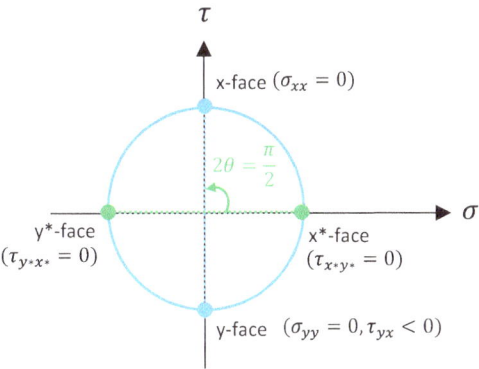

Going back to our tensor transformation formula, let us first compute the direction cosine matrix. Note that by convention, the first subscript refers to the new frame and the second subscript refers to the old frame. Therefore, the component "a_{xy}" represents the cosine of the angle between the new x-axis and the old y-axis. By observing simple geometry, we know that this angle is $\left(\frac{\pi}{2} + \theta\right) = \left(\frac{\pi}{2} + \frac{\pi}{4}\right)$ for our problem. Repeating this method, we can compute all values in the matrix as shown below.

$$a_{ik}a_{jl} = a_{ij} = \begin{vmatrix} a_{xx} & a_{xy} & a_{xz} \\ a_{yx} & a_{yy} & a_{yz} \\ a_{zx} & a_{zy} & a_{zz} \end{vmatrix} = \begin{vmatrix} \cos\dfrac{\pi}{4} & \cos\left(\dfrac{\pi}{2}+\dfrac{\pi}{4}\right) & 0 \\ \cos\left(\dfrac{\pi}{2}-\dfrac{\pi}{4}\right) & \cos\dfrac{\pi}{4} & 0 \\ 0 & 0 & \cos 0 \end{vmatrix}$$

$$= \begin{vmatrix} \dfrac{\sqrt{2}}{2} & -\dfrac{\sqrt{2}}{2} & 0 \\ \dfrac{\sqrt{2}}{2} & \dfrac{\sqrt{2}}{2} & 0 \\ 0 & 0 & 1 \end{vmatrix}$$

We need to show that the normal stress difference in the lab (new) frame is zero. Therefore, using the tensor transformation formula, we need to first find expressions for $\sigma^{new}{}_{xx}$ and $\sigma^{new}{}_{yy}$. We note that the old frame is also the principal frame; therefore, shear stresses are zero and only normal stresses $\sigma^{old}{}_{xx}$ and $\sigma^{old}{}_{yy}$ are non-zero. Therefore, our tensor transformation expressions may be simplified as shown below.

$$\sigma^{new}{}_{xx} = a_{xk}a_{xl}\sigma^{old}{}_{kl}$$
$$= a_{xx}a_{xx}\sigma^{old}{}_{xx} + a_{xx}a_{xy}\sigma^{old}{}_{xy} + a_{xy}a_{xx}\sigma^{old}{}_{yx} + a_{xy}a_{xy}\sigma^{old}{}_{yy} + a_{xz}a_{xz}\sigma^{old}{}_{zz}$$

$$\sigma^{new}{}_{xx} = \frac{\sqrt{2}}{2}\left(\frac{\sqrt{2}}{2}\right)\sigma^{old}{}_{xx} + 0 + 0 + \left(-\frac{\sqrt{2}}{2}\right)\left(-\frac{\sqrt{2}}{2}\right)\sigma^{old}{}_{yy} + 0$$

$$= \frac{1}{2}\sigma^{old}{}_{xx} + \frac{1}{2}\sigma^{old}{}_{yy}$$

We repeat the same process to find an expression for $\sigma^{new}{}_{yy}$:

$$\sigma^{new}{}_{yy} = a_{yk}a_{yl}\sigma^{old}{}_{kl}$$
$$= a_{yx}a_{yx}\sigma^{old}{}_{xx} + a_{yx}a_{yy}\sigma^{old}{}_{xy} + a_{yy}a_{yx}\sigma^{old}{}_{yx} + a_{yy}a_{yy}\sigma^{old}{}_{yy} + a_{yz}a_{yz}\sigma^{old}{}_{zz}$$

$$\sigma^{new}{}_{yy} = \frac{\sqrt{2}}{2}\left(\frac{\sqrt{2}}{2}\right)\sigma^{old}{}_{xx} + 0 + 0 + \frac{\sqrt{2}}{2}\left(\frac{\sqrt{2}}{2}\right)\sigma^{old}{}_{yy} + 0 = \frac{1}{2}\sigma^{old}{}_{xx} + \frac{1}{2}\sigma^{old}{}_{yy}$$

Therefore, the normal stress difference in the lab frame is shown to be zero.

$$\sigma^{new}{}_{xx} - \sigma^{new}{}_{yy} = \left(\frac{1}{2}\sigma^{old}{}_{xx} + \frac{1}{2}\sigma^{old}{}_{yy}\right) - \left(\frac{1}{2}\sigma^{old}{}_{xx} + \frac{1}{2}\sigma^{old}{}_{yy}\right) = 0$$

(d) We are told that the principal stress difference has a magnitude of 180 Pa, at a shear rate of 4 s^{-1}. We need to relate this information to viscosity of the fluid.

Therefore, we can start with the basic constitutive equation that relates shear stress to viscosity and shear rate as shown below.

$$\tau = \eta \dot{\gamma}$$

We note that our fluid flows in the direction of positive x-axis in the lab frame, this means that the opposing shear force acting on the fluid element will be in the negative x-axis direction and acting on the y-plane, this can be represented as τ_{yx}, and we observe that this direction is consistent with a negative value for τ_{yx} in the Mohr circle.

$$\left| \tau_{yx} \right| = \eta \dot{\gamma}$$

We note from our Mohr's circle construction earlier that $\sigma^*_{xx} > 0$ and $\sigma^*_{yy} < 0$; therefore the principal stress difference can be expressed as follows.

$$\sigma^*_{xx} - \sigma^*_{yy} = 180$$

By definition, the components of the stress tensor matrix that lie along the diagonal correspond to normal stresses, and the rest of the components correspond to shear stresses. Therefore, the following expression may be established.

$$\sigma = \begin{vmatrix} \sigma_{xx} & \sigma_{xy} & \sigma_{xz} \\ \sigma_{yx} & \sigma_{yy} & \sigma_{yz} \\ \sigma_{zx} & \sigma_{zy} & \sigma_{zz} \end{vmatrix} = \begin{vmatrix} \sigma_{xx} & \tau_{xy} & \tau_{xz} \\ \tau_{yx} & \sigma_{yy} & \tau_{yz} \\ \tau_{zx} & \tau_{zy} & \sigma_{zz} \end{vmatrix}$$

Now we apply the tensor transformation formula to find out σ^{new}_{yx} since it is equivalent to τ_{yx} in the lab frame which we are interested in to find out fluid viscosity. Again, we note that the old frame is also the principal frame; therefore, shear stresses are zero and only normal stresses σ^{old}_{xx} and σ^{old}_{yy} are non-zero. This helps us simplify the mathematical steps as shown below.

$$\sigma^{new}_{yx} = a_{yk} a_{xl} \sigma^{old}_{kl}$$
$$= a_{yx} a_{xx} \sigma^{old}_{xx} + a_{yx} a_{xy} \sigma^{old}_{xy} + a_{yy} a_{xx} \sigma^{old}_{yx} + a_{yy} a_{xy} \sigma^{old}_{yy} + a_{yz} a_{xz} \sigma^{old}_{zz}$$

$$\sigma^{new}_{yx} = \frac{\sqrt{2}}{2} \left(\frac{\sqrt{2}}{2} \right) \sigma^{old}_{xx} + 0 + 0 + \left(\frac{\sqrt{2}}{2} \right) \left(-\frac{\sqrt{2}}{2} \right) \sigma^{old}_{yy} + 0$$

$$\tau_{yx} = \frac{1}{2} \left(\sigma^{old}_{xx} - \sigma^{old}_{yy} \right) = \frac{1}{2} \left(\sigma^*_{xx} - \sigma^*_{yy} \right)$$

$$\left| \tau_{yx} \right| = \frac{1}{2}(180) = 90 \text{ Pa}$$

$$\eta = \frac{\left| \tau_{yx} \right|}{\dot{\gamma}} = \frac{90}{4} = 22.5 \text{ Pa.s}$$

(e) (i) We now have a polymer added to the glucose solution. The principal stress difference is increased to 2.3 times the original value, therefore,

$$\sigma^*_{xx} - \sigma^*_{yy} = 2.3(180) = 414 \text{ Pa}$$

In addition, the direction of the principal stress is now reduced to 25° to the direction of fluid flow; therefore, our direction cosine matrix is changed as follows:

$$a_{ik}a_{jl} = a_{ij} = \begin{vmatrix} a_{xx} & a_{xy} & a_{xz} \\ a_{yx} & a_{yy} & a_{yz} \\ a_{zx} & a_{zy} & a_{zz} \end{vmatrix} = \begin{vmatrix} \cos 25^\circ & \cos(90^\circ + 25^\circ) & 0 \\ \cos(90^\circ - 25^\circ) & \cos 25^\circ & 0 \\ 0 & 0 & \cos 0^\circ \end{vmatrix}$$

$$= \begin{vmatrix} 0.906 & -0.423 & 0 \\ 0.423 & 0.906 & 0 \\ 0 & 0 & 1 \end{vmatrix}$$

Now the normal stress difference in the lab frame can be computed in a similar way as the earlier part as follows:

$$\sigma^{new}_{xx} = a_{xk}a_{xl}\sigma^{old}_{kl}$$
$$= a_{xx}a_{xx}\sigma^{old}_{xx} + a_{xx}a_{xy}\sigma^{old}_{xy} + a_{xy}a_{xx}\sigma^{old}_{yx} + a_{xy}a_{xy}\sigma^{old}_{yy} + a_{xz}a_{xz}\sigma^{old}_{zz}$$

$$\sigma^{new}_{xx} = 0.906(0.906)\sigma^{old}_{xx} + 0 + 0 + (-0.423)(-0.423)\sigma^{old}_{yy} + 0$$
$$= 0.906^2\sigma^{old}_{xx} + 0.423^2\sigma^{old}_{yy}$$

$$\sigma^{new}_{yy} = a_{yk}a_{yl}\sigma^{old}_{kl}$$
$$= a_{yx}a_{yx}\sigma^{old}_{xx} + a_{yx}a_{yy}\sigma^{old}_{xy} + a_{yy}a_{yx}\sigma^{old}_{yx} + a_{yy}a_{yy}\sigma^{old}_{yy} + a_{yz}a_{yz}\sigma^{old}_{zz}$$

$$\sigma^{new}_{yy} = 0.423(0.423)\sigma^{old}_{xx} + 0 + 0 + 0.906(0.906)\sigma^{old}_{yy} + 0$$
$$= 0.423^2\sigma^{old}_{xx} + 0.906^2\sigma^{old}_{yy}$$

Therefore, the normal stress difference in the lab frame is now non-zero with a value that can be computed as shown below.

$$\sigma^{new}{}_{xx} - \sigma^{new}{}_{yy} = \left(0.906^2 \sigma^{old}{}_{xx} + 0.423^2 \sigma^{old}{}_{yy}\right) - \left(0.423^2 \sigma^{old}{}_{xx} + 0.906^2 \sigma^{old}{}_{yy}\right)$$

$$\sigma^{new}{}_{xx} - \sigma^{new}{}_{yy} = \left(0.906^2 - 0.423^2\right)\sigma^{old}{}_{xx} - \left(0.906^2 - 0.423^2\right)\sigma^{old}{}_{yy}$$

$$= 0.642 \left(\sigma^{old}{}_{xx} - \sigma^{old}{}_{yy}\right)$$

We know that $\sigma^*{}_{xx} - \sigma^*{}_{yy} = \sigma^{old}{}_{xx} - \sigma^{old}{}_{yy} = 414$ Pa, therefore,

$$\sigma^{new}{}_{xx} - \sigma^{new}{}_{yy} = 0.642(414) = 266 \text{ Pa}$$

Note that when polymer is added, the angle between the principal frame and the lab frame changes, and we no longer have a non-zero normal stress difference. It then follows that we are no longer in pure shear (Note that the corresponding Mohr's circle will no longer have its center at the origin).

(e) (ii) We can now compute shear stress and fluid viscosity in the same way as we did for the earlier part, since we know that shear rate is 4 s^{-1},

$$\sigma^{new}{}_{yx} = a_{yk} a_{xl} \sigma^{old}{}_{kl}$$
$$= a_{yx} a_{xx} \sigma^{old}{}_{xx} + a_{yx} a_{xy} \sigma^{old}{}_{xy} + a_{yy} a_{xx} \sigma^{old}{}_{yx} + a_{yy} a_{xy} \sigma^{old}{}_{yy} + a_{yz} a_{xz} \sigma^{old}{}_{zz}$$

$$\sigma^{new}{}_{yx} = \tau_{yx} = 0.423(0.906)\sigma^{old}{}_{xx} + 0 + 0 + 0.906(-0.423)\sigma^{old}{}_{yy} + 0$$

$$\tau_{yx} = 0.383\left(\sigma^{old}{}_{xx} - \sigma^{old}{}_{yy}\right) = 0.383\left(\sigma^*{}_{xx} - \sigma^*{}_{yy}\right)$$

$$\left|\tau_{yx}\right| = 0.383(414) = 158.6 \text{ Pa}$$

$$\eta = \frac{\left|\tau_{yx}\right|}{\dot{\gamma}} = \frac{158.6}{4} = 39.6 \text{ Pa.s}$$

The fluid viscosity here is also known as "apparent viscosity" as it is no longer Newtonian with a polymer added to it. Fluid viscosity is increased to a value of 39.6 Pa.s. The shear stress experienced by the fluid is also higher at 159 Pa. We can better analyze the changes brought about by the polymer addition by plotting the graph of apparent viscosity η_{app} against shear rate $\dot{\gamma}$. Assuming that the non-Newtonian behavior brought about by the polymer is shear-thinning (i.e., viscosity decreases with increasing shear rate), we arrive at the plot below.

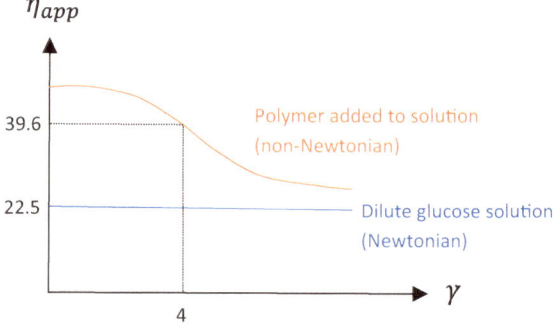

The dilute glucose solution is represented by a horizontal line as it is Newtonian and therefore viscosity is constant and independent of shear rate. When we add polymer to the solution, the viscosity at the same shear rate is increased, and shear-thinning with increasing shear rate causes the downward slope toward the right. Shear-thinning is commonly observed for cases when polymer is added and can be attributed to the viscoelasticity of the polymer introduced into the system.

(f) Examining stress values and stress differences when applying tensor transformation between coordinate frames (e.g., between principal and lab frames) can help with the analysis of flow characteristics. The stress difference can be used to compute shear stress and fluid viscosity in shear flow. This is useful as the results help us observe trends in viscosity and shear stress under different operating conditions. The understanding of fluid behavior under varying conditions helps us design and optimize processing parameters and fluid compositions to achieve desired outcomes.

Stress Difference and Tensor Transformation

In this problem, we note that the normal stress difference in the lab frame changed from zero to non-zero when the fluid properties changed from Newtonian (i.e., dilute glucose solution) to non-Newtonian (polymer added to solution).

Newtonian fluids are generally low in molecular weight with negligible elastic properties and do not display notable shear-thinning (viscosity remains constant). Examples of such fluids include water and very dilute solutions. Certain fluids may exhibit Newtonian behavior but only up to a certain point. If certain conditions, e.g., solution concentration, change and exceed the upper limit of the Newtonian region, then the fluid may display non-Newtonian behavior at those higher concentrations.

Non-Newtonian behavior can come about from viscoelastic properties, molecular rearrangements and structural changes, shear-thinning properties, or other triggering events such as nucleation and crystallization in concentrated solutions. Most fluids encountered in processing applications are in fact non-Newtonian, and some examples include viscoelastic polymeric systems which are commonly also shear-thinning. Non-polymeric suspensions are less elastic than polymeric systems but may still exhibit shear-thinning. Concentrated solutions may become unstable beyond certain concentration limits and start exhibiting non-Newtonian characteristics. Other non-Newtonian fluids include Boger fluids which are highly dilute and highly elastic fluids with constant viscosity (or only very weakly shear-thinning such that it is negligible). Boger fluids flow like liquid but behave like an elastic solid when subject to tensile forces and are made by adding a small amount of polymer to a Newtonian fluid with a high viscosity.

Sugar Solutions in Practical Applications

In practical applications, the study of sugary solutions may be of interest particularly in the food formulation and process design aspects of the confectionary industry. Food scientists are often interested in studying sugary solutions as they are commonly used as preservatives or ingredients. The normal stress difference can be related to viscosity, which is a good tell-tale sign of any non-Newtonian characteristics of interest. Experiments can be designed to examine viscosity patterns under varying sugar solution concentrations, temperatures, or with the presence of additives such as polymers. Understanding such trends will help better control the desired "thickness" (or viscosity) of the end product.

Polymer Addition

In the example of polymer addition to dilute solutions, one can intentionally alter a Newtonian-like dilute solution to non-Newtonian as the added polymer introduces viscoelasticity and shear-thinning properties which may be desired for ease of processing and/or enhanced food taste.

Solution Concentration

Other than polymer addition, solution concentration can also significantly affect product quality, one example being sucrose solution which displays Newtonian behavior at dilute concentrations but becomes less stable at high concentrations. Shearing flow causes collision between molecules which may trigger nucleation and faster crystal growth. When crystallization occurs, the characteristics of the solution can be markedly changed. Sucrose solutions have also been found to be more viscous at higher concentrations and start to transition from a free-flowing fluid state to a rubber-like state beyond certain concentration limits.

Problem 18

A stress matrix is typically used to store information related to the set of stress forces acting on a fluid particle. It is commonly written as a 3 by 3 matrix and contains both normal stresses and shear stresses.

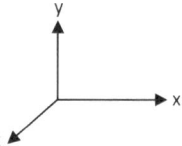

Describe the main types of deformation of a fluid particle, and assuming the coordinate system as shown above, illustrate the stresses acting on a unit cube of fluid, and relate the stress forces to the stress matrix.

Solution 18

Worked Solution

There are two main methods to deform a fluid particle:

1. Uniaxial extension
2. Stretching along its edges (shear)

Consider a unit cube of fluid. In the first case, extensional tensile stresses act to "pull" the fluid block along its long axis, while in the second case, shear stresses deform the fluid block by "stretching" it along its surfaces. Depending on the direction of these stresses with respect to the plane on which it acts, they may also be classified as a normal stress (i.e., perpendicular to surface) or shear stress (parallel to surface).

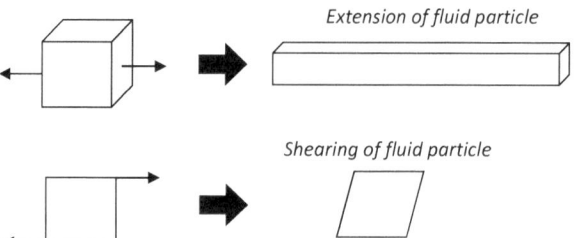

The stress matrix is defined by six stress forces, which comprise of normal and shear stresses. They are denoted with two subscripts whereby the first subscript defines the plane onto which the shear force acts. In other words, the direction of the normal vector to the plane determines the first subscript. The second subscript refers to the direction of the shear force itself (nothing to do with the plane onto which it acts). Therefore it is intuitive that the first and second subscripts are in fact the same for normal stresses (either x or y or z in an x-y-z coordinate system), and hence they adopt the common conventions in the terms σ_{xx}, σ_{yy}, and σ_{zz}.

As for shear stresses, the first and second subscripts represent directions that are perpendicular to each other, hence giving rise to the term τ_{xy}, whereby the x and y directions are at right angles to each other.

Putting together all shear forces acting on a unit cube of fluid, we can construct a stress matrix with nine elements. The diagram below shows the three stresses acting on the x-face (shaded below).

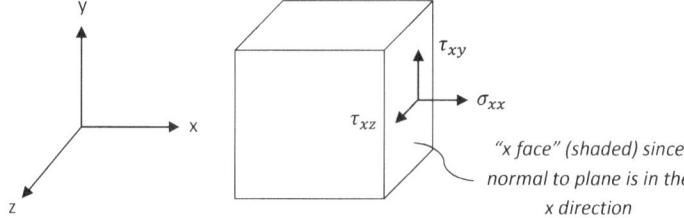

Following the same for the y and z planes, we are able to deduce the stresses σ_{yy}, τ_{yx}, and τ_{yz} on the y-face and the stresses σ_{zz}, τ_{zx}, and τ_{zy} on the z-plane. The results are combined in the 3 by 3 stress matrix shown below.

$$\begin{vmatrix} \sigma_{xx} & \tau_{xy} & \tau_{xz} \\ \tau_{yx} & \sigma_{yy} & \tau_{yz} \\ \tau_{zx} & \tau_{zy} & \sigma_{zz} \end{vmatrix}$$

Analyzing Fluid Flow Scenarios

Problem 19

Consider fluid flow in a channel with a rectangular cross section. The flow may be assumed to be a two-dimensional flow at steady state in the laminar regime. The length and height (over which the velocity profile develops) of the channel are 0.6 m and 5 mm, respectively. The width of the channel is 45 mm. Consider two possible ways in which the fluid flow can be modelled, assuming it behaves as a Newtonian fluid and power law fluid.

You may assume the following for the two cases:

- Newtonian: $\eta = 1.2$ Ns m^{-2}
- Power law: $k = 38$ Ns$^{0.8}$ m^{-2}, $n = 0.23$

Find the velocity profiles for the two cases. Then, assuming a pressure drop of 0.2 bar along the length of the channel, find the time required for "breakthrough" of a tracer and the mean residence time of the fluid for the two cases. [Breakthrough is defined as the occurrence when the first part of the tracer front reaches the end of the channel.]

© Springer Nature Switzerland AG 2019
X. W. Ng, *Pocket Guide to Rheology: A Concise Overview and Test Prep for Engineering Students*, https://doi.org/10.1007/978-3-030-30585-7_4

Solution 19

Worked Solution

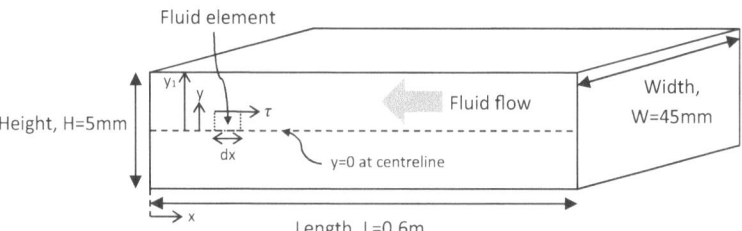

Newtonian Fluid

For a Newtonian fluid, shear stress is related to strain rate as follows, where viscosity η is a constant. Let v_x denote the fluid velocity in the positive x direction. We know that $\Delta P = 0.2$ bar.

$$\tau = \eta \dot{\gamma} = \eta \frac{dv_x}{dy}$$

We can also express shear stress in terms of pressure drop as follows.

$$\tau dx = y \frac{dP}{dx} dx$$

$$\tau = y \frac{dP}{dx} = y \frac{\Delta P}{L}$$

Equating the two expressions for τ, we get

$$y \frac{\Delta P}{L} = \eta \frac{dv_x}{dy}$$

$$\frac{\Delta P}{\eta L}(y)dy = dv_x$$

$$\frac{\Delta P}{\eta L}\left[\frac{y^2}{2}\right]_{y_1}^{y} = v_x - v_x|_{\text{wall}}$$

Due to the no-slip boundary condition at the duct wall, $v_x|_{\text{wall}} = 0$ when $y = y_1$; therefore, we obtain the velocity profile as follows.

$$\frac{\Delta P}{2\eta L}\left(y^2 - y_1{}^2\right) = v_x$$

To find "breakthrough" time, we need to first locate the point where maximum velocity occurs as the velocity profile is parabolic. This occurs at the point where $y = 0$.

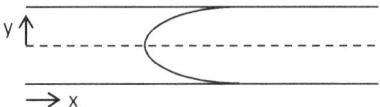

Substituting $y = 0$ into the velocity profile, we have

$$\frac{\Delta P}{2\eta L}\left(-y_1{}^2\right) = v_{x,\,max}$$

$$v_{x,\,max} = -\left(\frac{5 \times 10^{-3}}{2}\right)^2 \left(\frac{0.2 \times 10^5}{2(1.2)(0.6)}\right) = -0.0868 \text{ m/s}$$

Therefore time taken for breakthrough is

$$t_b = \frac{L}{|v_{x,\,max}|} = \frac{0.6}{0.0868} = 6.91 \text{ s}$$

In order to find the mean residence time, we will now need to find the mean velocity across the height of the duct. We can equate volumetric flow rates as follows:

$$2W\int_0^{y_1} v_x dy = 2y_1 W(v_{x,mean})$$

$$\int_0^{y_1} v_x dy = y_1(v_{x,mean})$$

$$v_{x,mean} = \frac{1}{y_1}\int_0^{y_1} v_x dy = \frac{1}{y_1}\int_0^{y_1}\left[\frac{\Delta P}{2\eta L}\left(y^2 - y_1{}^2\right)\right]dy$$

$$v_{x,mean} = \frac{\Delta P}{2y_1\eta L}\int_0^{y_1}\left(y^2 - y_1{}^2\right)dy$$

$$= \frac{\Delta P}{2y_1\eta L}\left[\frac{y^3}{3} - y_1{}^2 y\right]_0^{y_1} = \frac{\Delta P}{2y_1\eta L}\left[\frac{y_1{}^3}{3} - y_1{}^3\right] = -\frac{\Delta P}{2y_1\eta L}\left[\frac{2y_1{}^3}{3}\right]$$

$$v_{x,\text{mean}} = -\frac{\Delta P}{3\eta L}\left(y_1{}^2\right) = -\frac{0.2 \times 10^5}{3(1.2)(0.6)}\left(\frac{5 \times 10^{-3}}{2}\right)^2 = -0.0579 \text{ m/s}$$

Therefore, mean residence time is

$$t_m = \frac{L}{|v_{x,\text{mean}}|} = \frac{0.6}{0.0579} = 10.36 \text{ s}$$

You may observe that for a parabolic velocity profile in the case of a Newtonian fluid, the mean velocity is two-thirds of the maximum velocity at the centerline.

Power Law Fluid

For a power law fluid, the dependence of shear stress on strain rate is such that strain rate is raised to a power, denoted as n in this case. If $n = 1$, we have a Newtonian fluid; if $n < 1$, we have a shear-thinning fluid; and if $n > 1$, we have a shear-thickening fluid. In addition, viscosity is not constant but is a function of strain rate; hence, we coin a term called apparent viscosity, whereby $\eta_{\text{app}} = \eta_{\text{app}}(\dot{\gamma})$.

$$\tau = k\dot{\gamma}^n$$

$$\tau = \eta_{\text{app}}\dot{\gamma}, \quad \text{where } \eta_{\text{app}} = k\dot{\gamma}^{n-1}$$

Once again, we try to relate shear stress to velocity gradient:

$$\tau = k\left(\frac{dv_x}{dy}\right)^n = y\frac{\Delta P}{L}$$

$$\frac{dv_x}{dy} = \left(\frac{\Delta P}{kL}\right)^{\frac{1}{n}}y^{\frac{1}{n}}$$

Applying the no-slip boundary condition, $v_x|_{\text{wall}} = 0$ when $y = y_1$, we have

$$\left(\frac{\Delta P}{kL}\right)^{\frac{1}{n}}\left[\left(\frac{n}{1+n}\right)y^{\frac{1+n}{n}}\right]_{y_1}^{y} = v_x - v_x|_{\text{wall}}$$

$$v_x = -\left(\frac{n}{1+n}\right)\left(\frac{\Delta P}{kL}\right)^{\frac{1}{n}}\left[y_1{}^{\frac{1+n}{n}} - y^{\frac{1+n}{n}}\right]$$

To find the breakthrough time, we first determine the maximum velocity which occurs at the centerline where $y = 0$.

$$v_{x,\max} = -\left(\frac{n}{1+n}\right)\left(\frac{\Delta P}{kL}\right)^{\frac{1}{n}}\left(y_1^{\frac{1+n}{n}}\right)$$

$$= -\left(\frac{0.23}{1+0.23}\right)\left(\frac{0.2\times10^5}{38(0.6)}\right)^{\frac{1}{0.23}}\left(\left(\frac{5\times10^{-3}}{2}\right)^{\frac{1+0.23}{0.23}}\right)$$

$$v_{x,\max} = -0.0142 \text{ m/s}$$

Therefore time taken for breakthrough is

$$t_b = \frac{L}{|v_{x,\max}|} = \frac{0.6}{0.0142} = 42.25 \text{ s}$$

$$v_{x,\text{mean}} = \frac{1}{y_1}\int_0^{y_1} v_x dy = -\frac{1}{y_1}\int_0^{y_1}\left(\frac{n}{1+n}\right)\left(\frac{\Delta P}{kL}\right)^{\frac{1}{n}}\left[y_1^{\frac{1+n}{n}} - y^{\frac{1+n}{n}}\right]dy$$

$$= -\frac{1}{y_1}\left(\frac{n}{1+n}\right)\left(\frac{\Delta P}{kL}\right)^{\frac{1}{n}}\left[y_1^{\frac{1+n}{n}}y - \left(\frac{n}{1+2n}\right)y^{\frac{1+2n}{n}}\right]_0^{y_1}$$

$$= -\frac{1}{y_1}\left(\frac{n}{1+n}\right)\left(\frac{\Delta P}{kL}\right)^{\frac{1}{n}}\left(y_1^{\frac{1+2n}{n}} - \left(\frac{n}{1+2n}\right)y_1^{\frac{1+2n}{n}}\right)$$

$$= -\left(\frac{n}{1+n}\right)\left(\frac{\Delta P}{kL}\right)^{\frac{1}{n}}\left(y_1^{\frac{1+n}{n}}\right)\left(\frac{1+n}{1+2n}\right)$$

$$= -\left(\frac{n}{1+2n}\right)\left(\frac{\Delta P}{kL}\right)^{\frac{1}{n}}\left(y_1^{\frac{1+n}{n}}\right)$$

$$v_{x,\text{mean}} = -\left(\frac{0.23}{1.46}\right)\left(\frac{0.2\times10^5}{38(0.6)}\right)^{\frac{1}{0.23}}\left(2.5\times10^{-3}\right)^{\frac{1.23}{0.23}} = -0.0120 \text{ m/s}$$

Therefore, mean residence time is

$$t_m = \frac{L}{|v_{x,\text{mean}}|} = \frac{0.6}{0.0120} = 50 \text{ s}$$

You may observe that for this power law fluid with $n < 1$, it displays shear-thinning behavior such that compared to a Newtonian fluid, viscosity is lower for a given strain rate. For a shear-thinning fluid, viscosity decreases with increasing strain rate.

The velocity profile of this power law fluid is also more "flattened out" near to the centerline, as compared to the parabolic profile for a Newtonian fluid. In other words, it displays a more plugged flow profile. It follows that the value of mean velocity is closer to the value of maximum velocity at the centerline for the power law fluid, than for the Newtonian fluid.

Problem 20

Bingham fluids are typically viscoelastic and commonly encountered in processing applications. Examples of such fluids include mayonnaise and ketchup. Viscoelasticity refers to the ability to exhibit both viscous and elastic properties at the same time. For Bingham fluids in particular, they are rigid at low stresses but flow as a viscous fluid at high stresses.

Consider a Bingham fluid with viscosity η and yield stress τ_y that flows in a channel of length L with a square cross section with each side measuring W. It may be assumed that the flow is at steady state and laminar. Given that there is a pressure difference of ΔP between the two ends of the channel, derive an expression for the mean velocity, $\bar{u}$, for the fluid. Also, sketch the velocity profile and shear stress profile for this fluid and identify any key characteristics. The distance measured from the centerline of the channel can be denoted y, as shown in the diagram below.

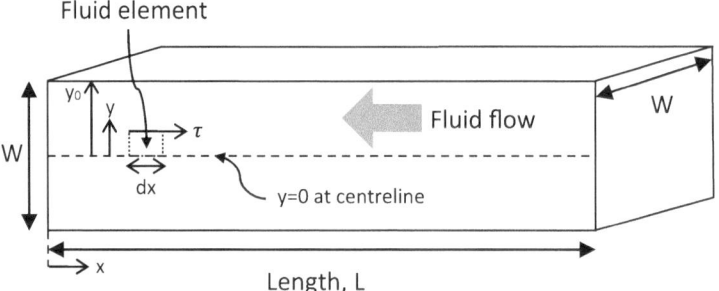

Solution 20

Worked Solution

For a non-Newtonian Bingham fluid, viscosity η is not a constant but varies with strain rate, $\eta = \eta(\dot{\gamma})$. Shear stress is related to strain rate as follows, where τ_y denotes yield stress and u denotes fluid velocity in the positive x direction.

$$\tau = \tau_y + \eta\dot{\gamma} = \tau_y + \eta\frac{du}{dy}$$

We can re-express shear stress in terms of pressure drop ΔP, which can be experimentally determined.

$$\tau = y\frac{dP}{dx} = y\frac{\Delta P}{L}$$

Combining the two expressions for τ, we get

$$y\frac{\Delta P}{L} = \tau_y + \eta\frac{du}{dy}$$

$$\frac{du}{dy} = y\frac{\Delta P}{\eta L} - \frac{\tau_y}{\eta}$$

We will now consider the two regimes that a Bingham fluid displays, first considering the case when shear stress is greater than yield stress. Hence for $\tau > \tau_y$, we have

$$u = \int_{y_0}^{y}\left(y\frac{\Delta P}{\eta L} - \frac{\tau_y}{\eta}\right)dy = \left(\frac{\Delta P}{\eta L}\right)\left[\frac{y^2}{2}\right]_{y_0}^{y} - \frac{\tau_y}{\eta}[y]_{y_0}^{y}$$

$$= -\left(\frac{\Delta P}{2\eta L}\right)\left[y_0{}^2 - y^2\right] + \frac{\tau_y}{\eta}[y_0 - y]$$

Now let's consider the case when $\tau < \tau_y$ within the inner core of the fluid. Shear rate decreases as we move further away from the channel walls and into the inner core. When shear rate just reaches zero, the value of shear stress corresponds to the yield stress τ_y. Therefore, we can find out the value of y at this point by substituting the condition $\dot{\gamma} = 0$ or $\frac{du}{dy} = 0$ as shown below.

$$\frac{du}{dy} = y\frac{\Delta P}{\eta L} - \frac{\tau_y}{\eta} = 0$$

Let y^* denote the value of y when yield stress is reached.

$$y^* = \frac{\tau_y L}{\Delta P}$$

The velocity at the value of yield stress can be expressed as follows, by substituting y^* into the velocity profile.

$$u^* = -\left(\frac{\Delta P}{2\eta L}\right)\left[y_0{}^2 - y^{*2}\right] + \frac{\tau_y}{\eta}[y_0 - y^*]$$

We can derive an expression for the mean velocity, $\bar{u}$, of the fluid by considering the total volumetric flow rate, Q, in the channel. Q is simply the sum of volumetric flow rates in two regions, the central core where the fluid flows with a constant velocity and the outer region where the fluid velocity varies with distance from the centerline.

$$Q = Q_{\text{core}} + Q_{\text{outer}}$$

$$W\bar{u}(2\,y_0) = Wu^*(2\,y^*) + W \int_{y^*}^{y_0} 2udy$$

$$\bar{u} = \frac{u^* y^*}{y_0} + \int_{y^*}^{y_0} udy$$

We can figure out the shear stress profile by noting that shear stress is the greatest near the walls and decreases to zero at the centerline due to symmetry. You may observe that the Bingham fluid has a velocity profile that has a plug flow shape in the central core and a more parabolic shape nearer the walls.

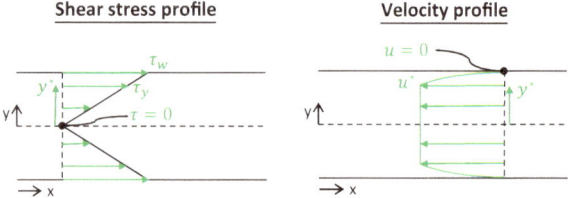

Problem 21

Constitutive equations may be applied to simple engineering flow geometries, such as pipe and capillary flow which are easy to achieve experimentally. Consider a Newtonian fluid under laminar flow whereby the Reynolds number Re is less than 500.

(a) Show that the following equation relating pressure drop and wall shear stress may be derived from a force balance.

$$\tau_w = \left(\frac{\Delta P}{L}\right)\frac{R}{2}$$

(b) Show that the pressure drop along the pipe can also be related to volumetric flow rate, Q, as shown below. Comment on factors to consider when sizing a pipe based on the equation below.

$$\Delta P = \frac{8L\eta Q}{\pi R^4}$$

(c) **Illustrate simple diagrams showing the variations of velocity, shear stress, and shear rate with respect to increasing radial length from the centerline of the pipe to the pipe wall.**

Solution 21

Worked Solution

(a) To recap, the Reynolds number is a dimensionless quantity that measures the ratio of inertial to viscous forces and therefore indicates the degree of laminar or turbulent flow. It may be expressed as follows whereby ρ represents fluid density; η and v represent dynamic and kinematic viscosities, respectively; u represents fluid velocity; and L represents the characteristic length scale for the specific flow geometry. This length would be pipe diameter for pipe flow, for example.

$$\mathrm{Re} = \frac{\rho u L}{\eta} = \frac{u L}{v}$$

For a low Reynolds number of less than 500, we may assume that flow is predominantly laminar and dominated by viscous force. In this regime, streamlines are parallel without significant mixing. The implicit assumption here is that any radial pressure changes are negligible $\frac{dP}{dr} = 0$. As Reynolds number increases to above about 2000, the flow starts to become turbulent as inertial forces dominate over viscous forces.

The force balance on a fluid element within a pipe flow can be better visualized with a diagram as shown below. Under steady laminar flow, the viscous force F_v contributed by shear stresses acting parallel to the streamlines balances the pressure force F_p contributed by the normal pressure force acting perpendicular to the cross section of the pipe.

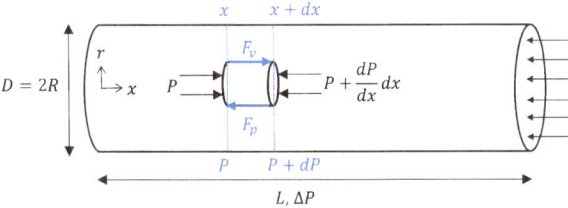

$$F_v - F_p = 0$$

The force balance equation may be expressed in terms of shear stress τ as shown below whereby A_s and A_c represent the surface area and cross-sectional area of the pipe, respectively, and P denotes the pressure force acting perpendicular to the cross section of the pipe.

$$\tau(A_s) - \left(\frac{dP}{dx}dx\right)A_c = 0$$

It is worth noting that the shear stress changes as we move along the radius of the pipe from its center. The shear stress at the centerline of the pipe where $r = 0$, is zero, $\tau = 0$. And shear stress increases until it reaches a maximum value $\tau = \tau_w$ at the wall of the pipe where $r = R$. This makes sense since there is maximum friction at the walls.

Going back to our force balance equation, we may simplify it as follows using our knowledge of circular geometry:

$$\tau(2\pi r dx) - \left(\frac{dP}{dx}dx\right)\pi r^2 = 0$$

$$\tau = \left(\frac{dP}{dx}\right)\frac{r}{2}$$

This differential equation can be adapted for the full length of the pipe as follows. We observe that shear stress varies linearly with radius (as measured from the centerline of the pipe).

$$\tau = \left(\frac{\Delta P}{L}\right)\frac{r}{2} \tag{1}$$

The maximum shear stress at the wall τ_w may also be derived by setting the condition of $r = R$. We therefore arrive at the given equation as shown below, which relates the pressure drop along the pipe length to the wall shear stress.

$$\tau_w = \left(\frac{\Delta P}{L}\right)\frac{R}{2}$$

(b) The constitutive equation for a Newtonian fluid can also be presented as follows, in terms of a constant fluid viscosity η and a shear rate denoted $\dot{\gamma}$. The magnitude of shear rate may then be expressed in terms of a velocity gradient.

$$\tau = \eta\dot{\gamma} = \eta\left|\frac{du}{dr}\right|$$

We may equate this expression for shear stress with the earlier equation (1) and integrate this differential equation with respect to radius r; this gives us the velocity profile for pipe flow as a function of r. Note that velocity decreases with increasing r; therefore, $\frac{du}{dr}$ is negative, and we need to add a negative sign to the $\frac{\Delta P}{L}$ term to ensure correctness in signs for this equation.

$$\eta \frac{du}{dr} = \left(-\frac{\Delta P}{L}\right)\frac{r}{2}$$

$$\frac{du}{dr} = -\frac{\Delta P}{2L\eta}r$$

$$u(r) = -\frac{\Delta P}{4L\eta}\left[r^2\right]_r^{r=R} = -\frac{\Delta P}{4L\eta}\left(R^2 - r^2\right)$$

We may observe from the form of the equation above that it is a quadratic function and hence describes a parabolic velocity profile. The volumetric flow rate Q can be easily obtained once we have determined the velocity function.

$$Q = \text{velocity} \times \text{cross} - \text{sectional area}$$

Since fluid velocity varies with radial length, we need to perform an integration over the pipe radius in order to find volumetric flow rate.

Differential ring element with cross-sectional area of $2\pi r dr$. Local velocity at this radial length r is $u(r)$

$$Q = \int_{r=0}^{r=R} u.2\pi r dr = \int_{r=0}^{r=R}\left[-\frac{\Delta P}{4L\eta}\left(R^2 - r^2\right)\right]2\pi r dr$$

$$Q = -\frac{\pi\Delta P}{2L\eta}\int_{r=0}^{r=R}\left(R^2 r - r^3\right)dr = -\frac{\pi\Delta P}{2L\eta}\left[R^2\frac{r^2}{2} - \frac{r^4}{4}\right]_0^R$$

$$Q = -\frac{\pi\Delta P}{2L\eta}\left[\left(\frac{R^4}{2} - \frac{R^4}{4}\right) - (0)\right] = -\frac{\pi R^4 \Delta P}{8L\eta}$$

$$\Delta P = -\frac{8L\eta Q}{\pi R^4}$$

Note here that the volumetric flow rate is a vector that has the same direction as the fluid velocity. Therefore, the negative sign means that the fluid flow direction is in the opposite direction as the direction of increasing pressure. This makes sense since the fluid flows down the pressure gradient. To simplify the analysis, we may decide to only deal with "magnitudes" here, i.e., the expression is written in terms of the magnitude (a positive value) of volumetric flow rate $|Q|$ as follows:

$$\Delta P = \frac{8L\eta|Q|}{\pi R^4}$$

We observe that the dependence of pressure drop along a circular pipe is non-linear ($\Delta P \sim R^{-4}$) with respect to the pipe radius. Understanding this correlation helps in pipe design to achieve desired flow conditions. This equation is useful since pressure drop along the pipe length ΔP and volumetric flow rate Q are easily measurable quantities, while the rest of the parameters in the equation can be easily obtained (i.e., pipe dimensions R, L). This equation allows us to determine viscosity η, which is a rheologically significant quantity.

We also notice that pressure drop is most sensitive to radius due to the dependence to the power of 4, as compared to other parameters such as viscosity and pipe length which vary linearly (to the power of 1) with pressure drop. It is important to understand factors leading to pressure drop especially due to practical considerations as there is a fine balance to strike in sizing pipes for operational needs. While it costs more to fabricate and maintain a bigger pipe (e.g., higher material cost), a larger pipe is ideal in terms of minimizing pressure drop which eventually saves on pumping cost.

(c) We have found earlier in part a the shear stress profile which has a linear dependence on r. We also determined the shear stress at the pipe wall.

$$\tau(r) = \left(\frac{\Delta P}{L}\right)\frac{r}{2}, \quad \tau_w = \left(\frac{\Delta P}{L}\right)\frac{R}{2} \tag{1}$$

We have also found earlier in part b the velocity profile for this flow which is parabolic with respect to r. We know that the velocity at the pipe wall is zero due to the no-slip boundary condition.

$$u(r) = -\frac{\Delta P}{4L\eta}\left(R^2 - r^2\right), \quad u_w = 0 \tag{2}$$

The remaining profile to be determined before we can plot the required profiles is the shear rate $\dot{\gamma}(r)$ as a function of pipe radius r. We know that the magnitude of shear rate is related to the magnitude of velocity gradient as follows:

$$\dot{\gamma} = \left|\frac{du}{dr}\right|$$

We can differentiate the velocity function $u(r)$ to obtain the required shear rate profile.

$$\dot{\gamma} = \left| \frac{du}{dr} \right| = \left| \frac{d}{dr} \left(-\frac{\Delta P}{4L\eta} \left(R^2 - r^2 \right) \right) \right| = \left| -\frac{\Delta P}{2L\eta} r \right| = \frac{\Delta P}{2L\eta} r$$

We can re-express shear rate in terms of the magnitude of volumetric flow rate $|Q|$ so that we may determine shear rate at the pipe wall in terms of easily measurable quantities $|Q|$ and pipe radius R.

$$\Delta P = \frac{8L\eta |Q|}{\pi R^4} \rightarrow \frac{\Delta P}{2L\eta} = \frac{4|Q|}{\pi R^4}$$

$$\dot{\gamma} = \frac{\Delta P}{2L\eta} r = \frac{4|Q|}{\pi R^4} r$$

At the pipe wall, the shear rate is therefore

$$\dot{\gamma}_w = \frac{4|Q|}{\pi R^4} R = \frac{4|Q|}{\pi R^3}$$

To summarize our results for shear rate, we have

$$\dot{\gamma}(r) = \frac{\Delta P}{2L\eta} r, \quad \dot{\gamma}_w = \frac{4|Q|}{\pi R^3} \tag{3}$$

With the results from Eqs. 1, 2, and 3, we arrive at the following plots for fluid flow from right to left. Note that the profiles are symmetrical about the centerline of the pipe. For a Newtonian fluid in laminar flow, we have a parabolic velocity profile and linear shear stress and shear rate profiles.

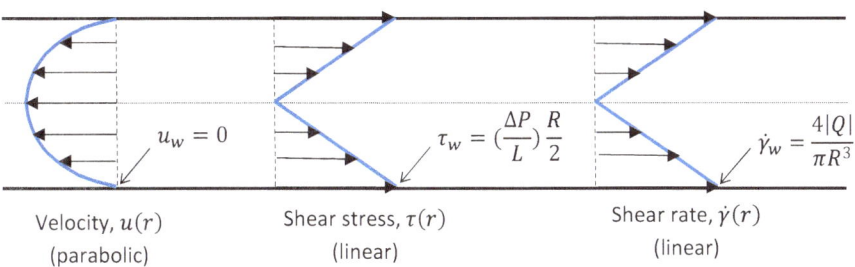

| Velocity, $u(r)$ | Shear stress, $\tau(r)$ | Shear rate, $\dot{\gamma}(r)$ |
| (parabolic) | (linear) | (linear) |

$u_w = 0$ $\tau_w = \left(\dfrac{\Delta P}{L} \right) \dfrac{R}{2}$ $\dot{\gamma}_w = \dfrac{4|Q|}{\pi R^3}$

Problem 22

Non-Newtonian constitutive equations may be used to deduce velocity profiles for shear-thinning and shear-thickening fluid flows in a cylindrical pipe. Using the power law model given below, and assuming steady laminar flow,

$$\tau = k\dot{\gamma}^n$$

(a) Show that the velocity profile for a non-Newtonian power law fluid flow may be expressed as follows:

$$u(r) = -\left(\frac{\Delta P}{2Lk}\right)^{1/n} \frac{n}{n+1}\left[R^{\frac{n+1}{n}} - r^{\frac{n+1}{n}}\right]$$

(b) Show that the pressure drop along the pipe ΔP may be expressed in terms of the magnitude of volumetric flow rate Q as follows:

$$\Delta P = 2Lk\left[\frac{3n+1}{\pi n R^{\frac{3n+1}{n}}}\right]^n |Q^n|$$

(c) Using your results from earlier parts, plot the velocity profiles for power law fluids with different values of n from $n < 1$ to $n = \infty$. Comment on your results.

Solution 22

Worked Solution

(a) The force balance on a fluid element in laminar flow can be done by equating the viscous force F_v contributed by shear stresses acting parallel to the flow direction and the pressure force F_p contributed by the normal pressure force acting perpendicular to the cross section of the pipe.

$$F_v = F_p$$

The force balance equation may be expressed in terms of shear stress τ as shown below, whereby ΔP is the pressure drop over the length of the small fluid element Δx:

$$\tau(2\pi r \Delta x) = \left(\frac{\Delta P}{L}\Delta x\right)\pi r^2$$

$$\tau = \left(\frac{\Delta P}{L}\right)\frac{r}{2}$$

We apply the constitutive equation for power law (non-Newtonian fluids) and integrate the differential equation as follows where C is an integration constant:

$$\tau = k\dot{\gamma}^n = k\left[\frac{du}{dr}\right]^n = \left(\frac{\Delta P}{L}\right)\frac{r}{2}$$

$$\frac{du}{dr} = \left[\left(\frac{\Delta P}{kL}\right)\frac{r}{2}\right]^{\frac{1}{n}} = \left(\frac{\Delta P}{2kL}\right)^{\frac{1}{n}}r^{\frac{1}{n}}$$

$$u(r) = \left(\frac{\Delta P}{2kL}\right)^{\frac{1}{n}}\left(\frac{n}{n+1}\right)r^{\frac{n+1}{n}} + C$$

Applying the boundary condition at the pipe wall, $r = R$, $u = 0$, we have

$$C = -\left(\frac{\Delta P}{2kL}\right)^{\frac{1}{n}}\left(\frac{n}{n+1}\right)R^{\frac{n+1}{n}}$$

Therefore, the velocity profile is as follows:

$$u(r) = -\left(\frac{\Delta P}{2kL}\right)^{\frac{1}{n}}\frac{n}{n+1}\left[R^{\frac{n+1}{n}} - r^{\frac{n+1}{n}}\right]$$

(b) Now we shall try to find out the relationship between shear stress τ and shear rate $\dot{\gamma}$ in order to plot the $\tau - \dot{\gamma}$ graph. We can differentiate the velocity function $u(r)$ to obtain shear rate which can then be related to shear stress via the constitutive equation for a power law fluid profile. Note that the negative sign means that the direction of shear rate $\dot{\gamma}$ is opposite to the direction of velocity $u(r)$.

$$\dot{\gamma} = \left|\frac{du}{dr}\right| = \frac{d}{dr}\left(\left(\frac{\Delta P}{2kL}\right)^{\frac{1}{n}}\frac{n}{n+1}\left[r^{\frac{n+1}{n}} - R^{\frac{n+1}{n}}\right]\right) = \left(\frac{\Delta P}{2kL}\right)^{\frac{1}{n}}r^{\frac{1}{n}}$$

$$Q = \int_{r=0}^{r=R} u \cdot 2\pi r dr = \int_{r=0}^{r=R}\left[-\left(\frac{\Delta P}{2kL}\right)^{\frac{1}{n}}\frac{n}{n+1}R^{\frac{n+1}{n}} - r^{\frac{n+1}{n}}\right]2\pi r dr$$

$$Q = -\frac{2\pi n}{n+1}\left(\frac{\Delta P}{2kL}\right)^{\frac{1}{n}}\int_{r=0}^{r=R}\left(rR^{\frac{n+1}{n}} - r^{\frac{2n+1}{n}}\right)dr$$

$$= -\frac{2\pi n}{n+1}\left(\frac{\Delta P}{2kL}\right)^{\frac{1}{n}}\left[\frac{r^2}{2}R^{\frac{n+1}{n}} - \frac{n}{3n+1}r^{\frac{3n+1}{n}}\right]_0^R$$

$$Q = -\frac{2\pi n}{n+1}\left(\frac{\Delta P}{2kL}\right)^{\frac{1}{n}}\left[\left(\frac{1}{2} - \frac{n}{3n+1}\right)\left(R^{\frac{3n+1}{n}}\right)\right]$$

$$= -\frac{\pi n}{n+1}\left(\frac{\Delta P}{2kL}\right)^{\frac{1}{n}}\left[\left(1 - \frac{2n}{3n+1}\right)\left(R^{\frac{3n+1}{n}}\right)\right]$$

$$Q = -\frac{\pi n}{n+1}\left(\frac{\Delta P}{2kL}\right)^{\frac{1}{n}}\left(\frac{n+1}{3n+1}\right)\left(R^{\frac{3n+1}{n}}\right) = -\pi\left(\frac{\Delta P}{2kL}\right)^{\frac{1}{n}}\left(\frac{n}{3n+1}\right)\left(R^{\frac{3n+1}{n}}\right)$$

$$Q^n = -\pi^n\left(\frac{\Delta P}{2kL}\right)\left(\frac{n}{3n+1}\right)^n\left(R^{\frac{3n+1}{n}}\right)^n$$

$$|Q^n| = \left(\frac{\Delta P}{2kL}\right)\left(\frac{\pi n R^{\frac{3n+1}{n}}}{3n+1}\right)^n$$

$$\Delta P = 2Lk\left[\frac{3n+1}{\pi n R^{\frac{3n+1}{n}}}\right]^n|Q^n|$$

(c) We observe from the velocity function found in part a that the velocity profile changes as the power law parameter n changes.

$$u(r) = -\left(\frac{\Delta P}{2kL}\right)^{\frac{1}{n}}\frac{n}{n+1}\left[R^{\frac{n+1}{n}} - r^{\frac{n+1}{n}}\right] = -\left(\frac{\Delta P}{2kL}\right)^{\frac{1}{n}}\frac{1}{1+\frac{1}{n}}\left[R^{1+\frac{1}{n}} - r^{1+\frac{1}{n}}\right]$$

For a base case of a Newtonian fluid, where we substitute $n = 1$ into the above equation, the velocity profile simplifies into a parabola as the function is quadratic in r as shown below.

$$u(r) \sim r^2$$

Shear-Thinning and Shear-Thickening Fluids

However, for a power law fluid that is non-Newtonian, we can have values of $n > 1$ for shear-thickening fluids and $n < 1$ for shear-thinning fluids. Upon examining the velocity function, we notice that as n increases from 1 to larger values, the exponent of r decreases from its value of 2 (the quadratic base case for a Newtonian fluid).

When taken to the upper limit where $n \to \infty$, $\frac{1}{n} \to 0$, the velocity profile becomes linear in r.

$$u(r) \sim r$$

   This means that as we transition from a Newtonian fluid into a more shear-thickening fluid, the velocity increases more slowly as we move from the pipe wall to the center of the pipe. Due to mass continuity (or mass conservation) along the pipe length under steady flow, a more gradual increase in velocity corresponds to an "extended velocity profile" (refer to diagram below) with a much larger peak velocity value reached. This makes sense as this allows the area under the velocity graph to remain the same regardless of type of fluid (shear-thickening/Newtonian, etc.) to satisfy mass continuity. Note however that the extreme case of $n \to \infty$ is physically unrealistic and is used here for analysis only.

   Conversely when we reduce the value of n from 1 to lower values, we are moving toward a shear-thinning fluid, and we observe from the mathematical equation that the exponent of r now increases to values larger than 2 (the quadratic base case for a Newtonian fluid). This means that the velocity rises very rapidly from zero at the pipe wall as we move away from the walls and achieves its maximum value quickly. This maximum value is maintained over most of the cross section of the pipe. The velocity profile can therefore be seen as more "compressed" and "flattened out" into a plateau near the central region of the pipe, with a lower maximum value reached. This type of plug flow profile is practically desirable due to the relative uniformity of velocities across the pipe cross section, hence achieving more consistent and pre-dictable flow characteristics.

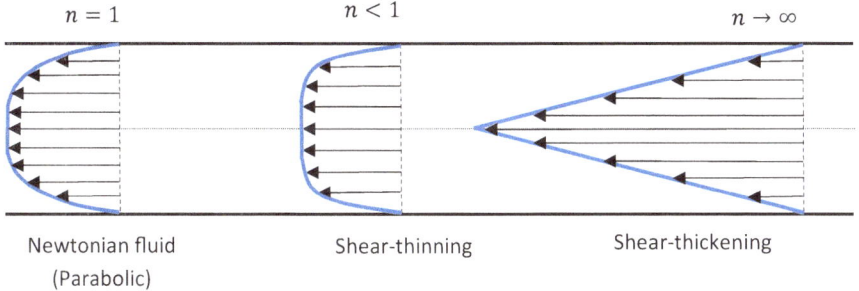

$n = 1$ $n < 1$ $n \to \infty$

Newtonian fluid
(Parabolic)   Shear-thinning   Shear-thickening

Problem 23

In industrial processing, residence time is a critical parameter that strongly affects product profile and overall yield achieved. For a material flowing through a volume, the residence time measures how much time the material spends in the volume. Therefore, an understanding of the residence time distribution (or RTD) provides an indication of reactor performance and fluid flow and helps in the design and optimization of processes.

In a particular residence time measurement, a tracer dye was added to two fluid samples A and B under steady laminar flow in a cylindrical pipe. The fluids were collected at the pipe exit over regular intervals, and the amount of tracer in the output samples was plotted. Comment on the results recorded below and describe the nature of fluids A and B in terms of:

- Velocity profiles in the direction of flow
- Breakthrough times
- Length of trailing end of RTD profile (i.e., "tail")
- Peak height of RTD profiles

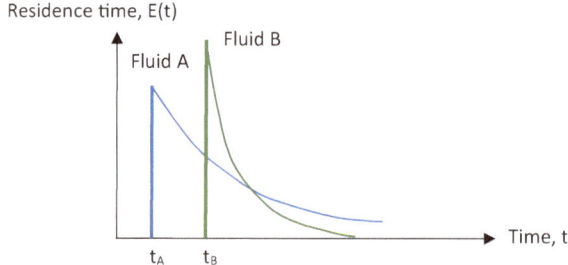

Solution 23

Worked Solution

The term breakthrough is used to describe the time when we first detect tracer in the output fluid. For the recorded results, this corresponds to the vertical lines observed at times t_A and t_B for fluid samples A and B, respectively.

We may summarize the observed results as follows for fluids A and B and explain the observations as shown:

	Fluid A	Fluid B
Velocity profile	Closer to parabolic profile	Closer to plug flow profile where it is "flattened out" near centerline of pipe
Breakthrough	Earlier breakthrough due to sharper peak exiting pipeline first	Later breakthrough ($t_B > t_A$) due to a shorter and flatter velocity profile. Velocity front (and peak of velocity at centerline) takes longer to exit the pipe
Peak height and "tail" of RTD	Lower peak and longer tail (slower decay) due to sharper velocity profile	Higher peak and shorter tail (faster decay) due to flatter velocity profile
Sketch of progression of velocity front		
Key characteristics	Typical parabolic profile of a Newtonian fluid which if modelled using the power law equation, $\tau = k\dot{\gamma}^n$ (where τ denotes shear stress, k and n are parametric constants, and $\dot{\gamma}$ denotes shear rate), has value of $n = 1$	Typical plug flow profile of a shear-thinning non-Newtonian fluid which if modelled using the power law equation, $\tau = k\dot{\gamma}^n$ (where τ denotes shear stress, k and n are parametric constants, and $\dot{\gamma}$ denotes shear rate), has value of $n < 1$

Problem 24

Consider Couette flow for a polymer melt in a narrow gap measuring 1.2 mm between two coaxial cylinders, whereby the inner cylinder is rotating, while the outer cylinder is held stationary. The diameter of the inner cylinder is 25 mm, while the lengths of the cylinders are the same at 90 mm.

(a) Describe key features of Couette flow and compare them with other possible flow characteristics.
(b) Given that for a particular flow condition, the direction of the principal axis is 30° to the direction of flow and points in the plane of shear of the flow, determine the shear stress using tensor notation if the principal stress difference is 25 kPa.
(c) Determine the torque measured on the inner cylinder.
(d) Given that the inner cylinder rotates at an angular velocity of 250 radians per second, calculate:

  (i) The apparent shear viscosity of the polymer melt
  (ii) The normal stress difference of the polymer melt in the laboratory frame

Solution 24

Worked Solution

(a) Couette flow is a type of drag-induced flow and can occur either between large parallel flat plates or between long concentric rotating cylinders (with each cylinder rotating at different speeds). This type of fluid motion driven by viscous drag force (or shear force) is commonly distinguished from pressure-driven flow, also known as Poiseuille flow, where an applied pressure gradient drives fluid motion. However, note that fluid flow can also be driven concurrently by both drag and an applied pressure gradient.

(b) Before we find torque, we need to first calculate shear stress τ. The magnitude of torque $|T|$ can then be determined by multiplying shear stress τ with the surface area over which it acts (i.e., curved surface area of inner cylinder of length L) and with the radius of the inner cylinder r_1 (i.e., distance perpendicular to direction of rotational flow).

$$|T| = |\tau| \times 2\pi r_1 L \times r_1$$

Let us first visualize the flow pattern in a narrow Couette gap measuring δ where the bottom surface is moving faster relative to the top surface, dragging the fluid in the gap along with velocity profile shown below in blue arrows. Consider also, the direction of the principal axis (x_1) at an angle of 30° to the direction of flow (x_1').

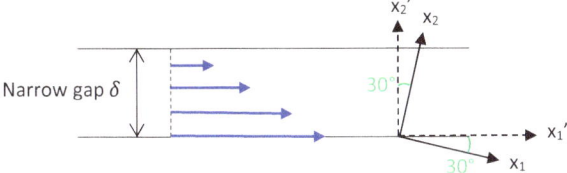

In the original coordinate frame in which our principal axes were defined, i.e., x_1 and x_2, the stress matrix σ can be expressed as follows. As there are no shear force components σ_{ij} in this frame, only the principal (or normal) stresses have non-zero values as indicated. We also note that $\sigma_{33} = 0$ as our flow condition is two-dimensional.

$$\sigma = \begin{vmatrix} \sigma_{11} & \sigma_{12} & \sigma_{13} \\ \sigma_{21} & \sigma_{22} & \sigma_{23} \\ \sigma_{31} & \sigma_{32} & \sigma_{33} \end{vmatrix} = \begin{vmatrix} \sigma_{11} & 0 & 0 \\ 0 & \sigma_{22} & 0 \\ 0 & 0 & 0 \end{vmatrix}$$

From the angle of rotation between the two frames, we can determine the directional cosine matrix.

$$\begin{vmatrix} a_{11} & a_{12} & a_{13} \\ a_{21} & a_{22} & a_{23} \\ a_{31} & a_{32} & a_{33} \end{vmatrix} = \begin{vmatrix} \cos 30° & \cos(90° - 30°) & 0 \\ \cos(90° + 30°) & \cos 30° & 0 \\ 0 & 0 & \cos 0° \end{vmatrix}$$

$$= \begin{vmatrix} 0.866 & 0.5 & 0 \\ -0.5 & 0.866 & 0 \\ 0 & 0 & 1 \end{vmatrix}$$

We can relate the stress matrix in the new frame σ' (or laboratory frame) to that in the old frame using the directional cosines using this general expression

$$\sigma'_{ij} = a_{ik}a_{jl}\sigma_{kl} = \cos\theta_{ik}\cos\theta_{jl}\sigma_{kl}$$

$$\sigma'_{ij} = \begin{vmatrix} \sigma'_{11} & \sigma'_{12} & \sigma'_{13} \\ \sigma'_{21} & \sigma'_{22} & \sigma'_{23} \\ \sigma'_{31} & \sigma'_{32} & \sigma'_{33} \end{vmatrix}$$

We know that since $a_{31} = a_{32} = a_{13} = a_{23} = 0$, therefore $\sigma'_{31} = \sigma'_{32} = \sigma'_{13} = \sigma'_{23} = 0$.

$$\sigma'_{ij} = \begin{vmatrix} \sigma'_{11} & \sigma'_{12} & 0 \\ \sigma'_{21} & \sigma'_{22} & 0 \\ 0 & 0 & \sigma'_{33} \end{vmatrix}$$

Now we can compute the value of shear stress τ which is in fact $\tau_{21} = \sigma'_{21}$ since it acts in the direction of x_1', in the x_2' plane. As our stress matrix is symmetric, $\tau_{21} = \sigma'_{21} = \tau_{12} = \sigma'_{12}$.

$$\tau = \tau_{12} = \sigma'_{12} = a_{1k}a_{2l}\sigma_{kl} = a_{11}a_{21}\sigma_{11} + a_{11}a_{22}\sigma_{12} + a_{12}a_{21}\sigma_{21} + a_{12}a_{22}\sigma_{22}$$
$$+ a_{13}a_{23}\sigma_{33}$$

We note from the values in the original stress matrix σ that only σ_{11} and σ_{22} are non-zero values. Therefore,

$$\tau = a_{11}a_{21}\sigma_{11} + a_{12}a_{22}\sigma_{22} = 0.866(-0.5)\sigma_{11} + 0.5(0.866)\sigma_{22}$$
$$\tau = -0.433\sigma_{11} + 0.433\sigma_{22} = -0.433(\sigma_{11} - \sigma_{22})$$

We were given the value of the principal stress difference to be 25 kPa; therefore, we can find our shear stress as follows.

$$\tau = -0.433(25000) = -10825 \text{ Pa}$$
$$|\tau| = 10825$$

(c) Substituting the value of shear stress back into our earlier expression for torque,

$$|T| = |\tau| \times 2\pi r_1 L \times r_1 = 10825 \times 2\pi(12.5 \times 10^{-3})(90 \times 10^{-3}) \times (12.5 \times 10^{-3})$$
$$= 0.956 \text{ Nm}$$

(d) (i) For a narrow gap, we can apply the lubrication approximation which simplifies calculations by assuming a constant strain rate $\dot{\gamma}$. By definition, strain rate can be expressed in terms of fluid velocity V and gap size δ as follows. Also, we know that for rotational flow, the angular velocity ω is related to tangential velocity V and distance from the axis of rotation (i.e., r_1 in this problem) according to the equation $V = r_1\omega$.

$$|\dot{\gamma}| = \left|\frac{V}{\delta}\right| = \frac{r_1\omega}{\delta} = \frac{(12.5 \times 10^{-3})(250)}{1.2 \times 10^{-3}} = 2604 \text{ } s^{-1}$$

Apparent shear viscosity n_{app} can now be found using strain rate $|\dot{\gamma}|$ and the shear stress value $|\tau|$ found earlier.

$$n_{app} = \left|\frac{\tau}{\dot{\gamma}}\right| = \frac{10825}{2604} = 4.16 \; Pas$$

(d) (ii) In order to calculate the normal stress difference in the laboratory frame, we need to compute the normal stress values in this frame.

$$\sigma'_{11} = a_{1k}a_{1l}\sigma_{kl} = a_{11}a_{11}\sigma_{11} + a_{11}a_{12}\sigma_{12} + a_{12}a_{11}\sigma_{21} + a_{12}a_{12}\sigma_{22} + a_{13}a_{13}\sigma_{33}$$

Same as before, we simplify the above calculation knowing that only σ_{11} and σ_{22} are non-zero values.

$$\sigma'_{11} = a_{11}a_{11}\sigma_{11} + a_{12}a_{12}\sigma_{22} = \left(0.866^2\right)\sigma_{11} + \left(0.5^2\right)\sigma_{22}$$

We can do the same to determine the value of σ'_{22} as follows.

$$\sigma'_{22} = a_{21}a_{21}\sigma_{11} + a_{22}a_{22}\sigma_{22} = \left((-0.5)^2\right)\sigma_{11} + \left(0.866^2\right)\sigma_{22}$$

$$= \left(0.5^2\right)\sigma_{11} + \left(0.866^2\right)\sigma_{22}$$

We can now determine the normal stress difference for the laboratory frame, $(\sigma'_{11} - \sigma'_{22})$ using the known value of normal stress difference in the principal frame $(\sigma_{11} - \sigma_{22})$ and the values calculated above.

$$\sigma'_{11} - \sigma'_{22} = \left(0.866^2\right)\sigma_{11} + \left(0.5^2\right)\sigma_{22} - \left(0.5^2\right)\sigma_{11} - \left(0.866^2\right)\sigma_{22}$$

$$\sigma'_{11} - \sigma'_{22} = (\sigma_{11} - \sigma_{22})\left(0.866^2 - 0.5^2\right) = (25000)(0.5) = 12500 \; Pa$$

Problem 25

Experimental data for a particular molten polymer at 220 °C was found to fit the following equation.

$$\tau(t) = -\int_{-\infty}^{t} 120e^{-(t-t')/8}\gamma(t,t')dt'$$

Given that a steady strain rate of $\dot{\gamma} = 8 \; s^{-1}$ was applied to the melt for a period of time, after which shearing was stopped at time $t = 0$,

(a) **Find the stress relaxation response after termination of shear.**

(b) **Plot the stress relaxation curve for the result in part a and indicate how a purely elastic and a purely viscous fluid are expected to behave.**

(c) **Consider instead that the polymer melt at 220 °C starts from rest before being subject to a constant strain rate of $\dot{\gamma} = 8\ s^{-1}$.**

  (i) **Determine the strain and stress growth profiles and sketch them.**
  (ii) **Sketch the stress growth profile if the temperatures were 240 °C and 200 °C instead.**

(d) **Discuss how the model can be altered to reflect a more realistic response for a typical commercial polydisperse viscoelastic melt with shear-thinning properties.**

Solution 25

Worked Solution

(a) Before we proceed, we should clarify our understanding of current time and past time. In the equation given in this problem as shown below, shear stress τ is a function of current time t, and t has a value of zero when shearing just stops. However prior to termination of shear, there was a constant strain rate. This is captured in the integral shown below where t' refers to a past time variable that can take values prior to $t = 0$ (shearing just stops), i.e., t' stretches back in time from $t' = t = 0$ to $t' = -\infty$.

$$\tau(t) = -\int_{-\infty}^{t} 120 e^{-(t-t')/8} \gamma(t, t') dt'$$

Strain γ can be written as follows, where its value is a cumulative effect (as expressed by the integral operation) over a particular period in history (i.e., before $t = 0$).

$$\gamma(t, t') = \int_{t}^{t'} \dot{\gamma}(t') dt'$$

Going back to the problem, we know that $\dot{\gamma} = 8\ s^{-1}$ for $t' < 0$ and $\dot{\gamma} = 0\ s^{-1}$ for $t' \geq 0$. Therefore we can express strain as follows:

$$\gamma(t, t') = \int_{0}^{t'} 8 dt' = 8t'$$

Substituting this expression back in to the shear stress function, we have

$$\tau(t) = -\int_{-\infty}^{0} 120e^{-\frac{(t-t')}{8}}(8t')dt' = -960\int_{-\infty}^{0} e^{-\frac{(t-t')}{8}}t'dt' = -960e^{-\frac{t}{8}}\int_{-\infty}^{0} e^{\frac{t'}{8}}t'dt'$$

In order to solve the integral, we can use integration by parts, which is defined as follows:

$$\int u\,dv = uv - \int v\,du$$

$$u = t' \rightarrow du = dt'$$

$$dv = e^{\frac{t'}{8}}dt' \rightarrow v = 8e^{\frac{t'}{8}}$$

$$\int_{-\infty}^{0} e^{\frac{t'}{8}}t'dt' = \left[8t'e^{\frac{t'}{8}}\right]_{-\infty}^{0} - \int_{-\infty}^{0} 8e^{\frac{t'}{8}}dt' = 0 - \left[64e^{\frac{t'}{8}}\right]_{-\infty}^{0} = -64$$

Substituting back into the shear stress expression, we obtain the stress relaxation response as a function of current time t. We observe that shear stress experiences an exponential decay upon cessation of shear.

$$\tau(t) = -960e^{-\frac{t}{8}}\int_{-\infty}^{0} e^{\frac{t'}{8}}t'dt' = -960e^{-\frac{t}{8}}(-64) = 61440e^{-\frac{t}{8}}$$

(b)

Purely Elastic

In a fully elastic system, there is reversible deformation where all energy is preserved. This system is described by Hooke's law and represents the elastic limit of viscoelasticity. For a pure elastic component that has no viscous component, there is no viscous dissipation, and shear stress is held at the same constant value that it had at $t = 0$ even after shearing has stopped for $t > 0$. This is shown by the blue dotted line in the plot below.

Purely Viscous

Conversely, in a fully viscous system, there is irreversible deformation and all energy is lost as heat. The final shape of the fluid is determined by the shape of the container. Viscosity is a measure of resistance to flow: therefore, a higher viscosity leads to greater energy dissipation. A fully viscous component has no elastic contributions and responds instantaneously to strain rate. This results in a step change to zero the moment shearing ceases, and this value remains zero for the period of zero shear as shown by the orange dotted line in the plot below. A fully viscous substance is said to be Newtonian with constant viscosity. Most substances with low molecular mass are Newtonian, and they can include gases, molten metals, organic and inorganic liquids, or solutions of inorganic salts.

Viscoelastic Fluid

The purely elastic and purely viscous states represent the two extreme theoretical limits, and they are more often the minority than the norm for most real fluids. Most real fluids of industrial importance exhibit viscoelastic behavior that lies somewhere between these two limits under different operating conditions. Some examples include multiphase systems such as foams, emulsions, slurries, suspensions and dispersions, and polymer melts and polymer solutions. These fluids are often called non-Newtonian and rheologically complex as they do not conform to the simplistic linear Newtonian behavior.

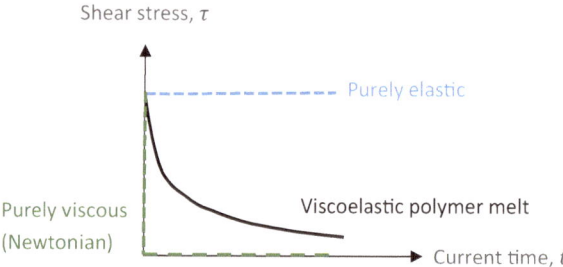

(c) (i) Now we have the polymer melt starting from rest instead of having a prior history of constant shear. Therefore, we have the following conditions:

$$\dot{\gamma} = 0s^{-1}, \quad t' < 0$$
$$\dot{\gamma} = 8s^{-1}, \quad t' \geq 0$$

Now note that strain is expressed as follows in its general form. Observe that the value of strain is non-zero even during the period $t' < 0$ when $\dot{\gamma} = 0s^{-1}$.

$$\gamma(t, t') = \int_{t}^{t'} \dot{\gamma} dt'$$

The strain function when $t' \geq 0$ and $\dot{\gamma} = 8$ is as follows:

$$\gamma(t, t') = \int_{t}^{t'} \dot{\gamma} dt' = 8(t' - t) = -8(t - t')$$

The strain function when $t' < 0$ and $\dot{\gamma} = 0$ is as follows:

$$\gamma(t, t') = \int_t^0 8dt' = 8(0 - t) = -8t$$

We can now sketch the required plot for strain as shown below:

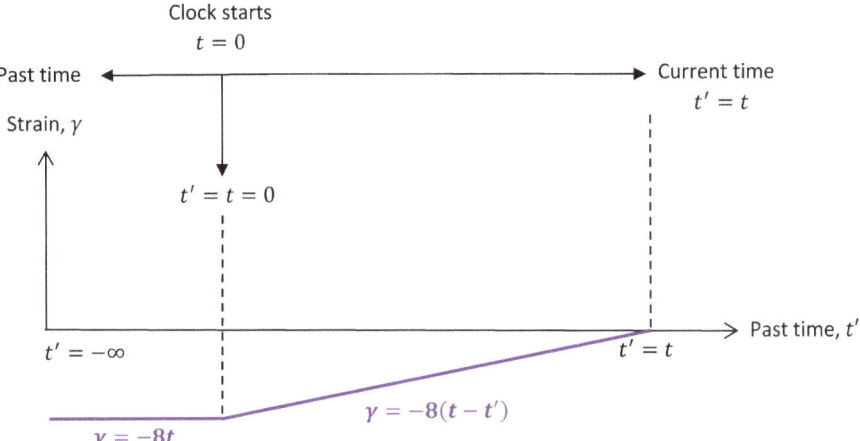

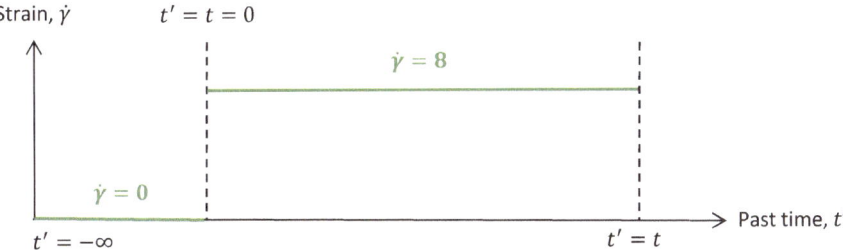

Now we can determine the stress growth profile. We need to also break up our integral for shear stress into two time periods, as they experience different strain profiles. Substituting the strain expressions found earlier into the stress function, we have

$$\tau(t) = -\int_{-\infty}^t 120e^{-\frac{(t-t')}{8}}\gamma(t, t')dt'$$

$$= -\int_{-\infty}^0 120e^{-\frac{(t-t')}{8}}\gamma(t, t')dt' - \int_0^t 120e^{-\frac{(t-t')}{8}}\gamma(t, t')dt'$$

$$\tau(t) = -120\int_{-\infty}^0 e^{-\frac{(t-t')}{8}}(-8t)dt' - \int_0^t 120e^{-\frac{(t-t')}{8}}(-8(t - t'))dt'$$

$$\tau(t) = 960 \int_{-\infty}^{0} e^{-\frac{(t-t')}{8}} t dt' - 960 \int_{0}^{t} e^{-\frac{(t-t')}{8}} (t' - t) dt'$$

Let us solve the integrals one step at a time. For the first integral, we note that the variable we are integrating with respect to is t' and not t. Therefore t is similar to a constant in this integral which can be taken out of the integral.

$$960 \int_{-\infty}^{0} e^{-\frac{(t-t')}{8}} t dt' = 960 t e^{-\frac{t}{8}} \int_{-\infty}^{0} e^{\frac{t'}{8}} dt' = 960 t e^{-\frac{t}{8}} \left[8 e^{\frac{t'}{8}} \right]_{-\infty}^{0} = 960 t e^{-\frac{t}{8}} [8 - 0]$$

$$= 7680 t e^{-\frac{t}{8}}$$

As for the second integral, we can use a substitution of variables to make the integration easier to handle, let $s = t - t'$. And it follows that $ds = -dt'$ and when $t' = 0$, $s = t$ and when $t' = t$, $s = 0$. Therefore,

$$960 \int_{0}^{t} e^{-\frac{(t-t')}{8}} (t' - t) dt' \rightarrow 960 \int_{t}^{0} e^{-\frac{s}{8}} (-s)(-ds) = 960 \int_{t}^{0} e^{-\frac{s}{8}} s ds$$

In order to solve this integral, we can use integration by parts, which is defined as follows:

$$\int u dv = uv - \int v du$$

$$u = s \rightarrow du = ds$$

$$dv = e^{-\frac{s}{8}} ds \rightarrow v = -8 e^{-\frac{s}{8}}$$

$$\int_{t}^{0} e^{-\frac{s}{8}} s ds = \left[-8 s e^{-\frac{s}{8}} \right]_{t}^{0} - \int_{t}^{0} -8 e^{-\frac{s}{8}} ds = 8 t e^{-\frac{t}{8}} - \left[64 e^{-\frac{s}{8}} \right]_{t}^{0}$$

$$= 8 t e^{-\frac{t}{8}} - \left[64 - 64 e^{-\frac{t}{8}} \right] = 8 t e^{-\frac{t}{8}} + 64 e^{-\frac{t}{8}} - 64$$

Putting our results together, we have determined the following stress growth profile.

$$\tau(t) = 960 \int_{-\infty}^{0} e^{-\frac{(t-t')}{8}} t dt' - 960 \int_{0}^{t} e^{-\frac{(t-t')}{8}} (t' - t) dt'$$

$$= 7680 t e^{-\frac{t}{8}} - 960 \left(8 t e^{-\frac{t}{8}} + 64 e^{-\frac{t}{8}} - 64 \right)$$

$$\tau(t) = 7680 t e^{-\frac{t}{8}} - 7680 t e^{-\frac{t}{8}} - 61440 e^{-\frac{t}{8}} + 61440 = 61440 \left(1 - e^{-\frac{t}{8}} \right)$$

This function describes an exponential stress growth, and its shape can be seen in the plot below. We can see that shear stress reaches an asymptotic value of 61,440 Pa at large times.

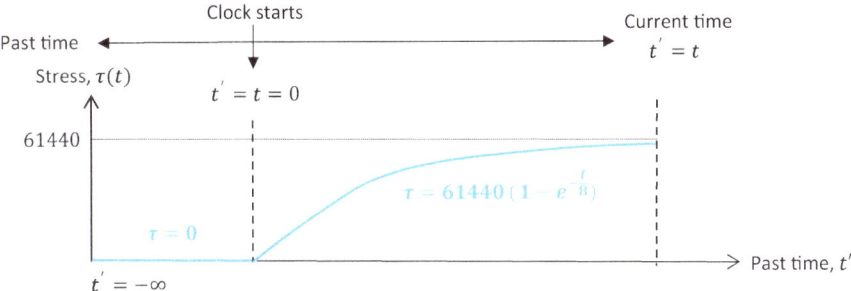

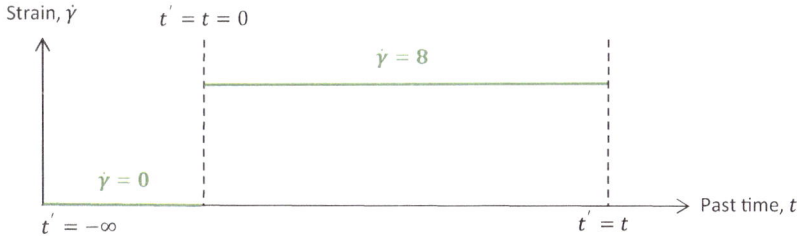

(c) (ii) This question checks our understanding on the temperature dependence of viscoelastic properties. For the elastic constant g, we note that in most cases it increases as temperature decreases. Since g is a ratio of stress to strain for an elastic component, it also quantifies "stiffness" where a more elastic material is stiffer.

$$ g = \frac{\tau}{\gamma} $$

It takes less effort to stretch a viscoelastic material for a specific distance at a higher temperature (material is less stiff) than at a lower temperature. This is also the reason why extremely low temperatures can cause some materials to stiffen and become brittle and break.

The other viscoelastic property, viscosity η, also depends on temperature according to an Arrhenius-type relationship, whereby viscosity increases as temperature decreases. In the equation below, η_0 denotes a pre-exponential factor, E denotes activation energy, and R is the gas constant.

$$\eta = \eta_0 e^{\frac{E}{RT}}$$

Therefore, we can deduce the following plots at a higher and lower temperature than the base case of 220 °C.

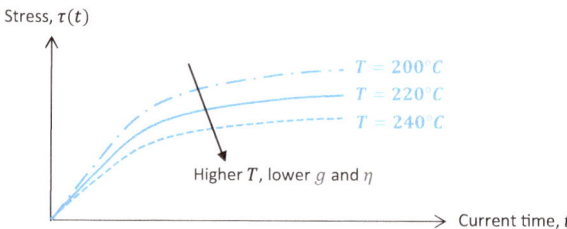

(d) The general form of the shear stress function is shown below, where in our problem we have relaxation time $\lambda = 8s$ and $\frac{g}{\lambda} = 120 \, Pa.s^{-1}$

$$\tau(t) = -\int_{-\infty}^{t} \frac{g}{\lambda} e^{-(t-t')/\lambda} \gamma(t, t') dt' = -\int_{-\infty}^{t} 120 e^{-(t-t')/8} \gamma(t, t') dt'$$

Most commercial polymers are polydisperse, which means that they comprise of a distribution of sizes (or molecular masses) instead of just a single uniform size (monodisperse). In order to modify our stress function to better represent polydisperse polymers, we can modify our function by including a summation over the range of timescales λ_i and elastic constants g_i, contributed by the distribution of polymers ($i = 1$ to N where N is the number of differently sized polymers present). This is also called the multimodal model and can be seen from the adjustment in green below.

$$\tau(t) = -\int_{-\infty}^{t} \sum_{i=1}^{N} \frac{g_i}{\lambda_i} e^{-(t-t')/\lambda_i} \gamma(t, t') dt'$$

Most commercial polydisperse polymer melts are also shear-thinning, which means that apparent viscosity decreases with increasing strain rate at a range of higher strain rate values. In other words, shear stress curve increases less rapidly due to the shear-thinning effect, especially in the range of higher strain rates. This characteristic can be accounted for by adjusting the model as shown in blue below, where we introduce a "damping term" experiencing exponential decay. This term introduces a strain dependence into the stress function and depresses the value of τ for a given strain, especially at higher strain rates. The factor k ranges from 0 to 1 depending on the severity of the shear-thinning effect. Some common substances that undergo shear-thinning are tomato ketchup, whipped cream, nail polish, and silicone oil.

$$\tau(t) = - \int_{-\infty}^{t} \frac{g}{\lambda} e^{-(t-t')/\lambda} \gamma(t,t') e^{-k|\gamma(t,t')|} dt'$$

Problem 26

The Cross equation as shown below is often used to model a non-Newtonian shear-thinning suspension under steady-state flow.

$$\eta_{\text{app}} - \eta_{\infty} = \frac{\eta_0 - \eta_{\infty}}{1 + \alpha \dot{\gamma}^m}$$

(a) **Show how the Cross equation above may be derived from first principles by considering interactions between particles in the suspension. You may assume that the number of interparticle interactions is linearly proportional to viscosity contribution by the particles.**

(b) **Plot a graph of viscosity η_{app} against shear rate $\dot{\gamma}$ and comment on its shape by considering what happens at the particle level.**

Solution 26

Worked Solution

(a) The Cross equation applies well to suspensions or polymer solutions whereby there are particles suspended in a base fluid. Changes in viscosity with varying shear rate are therefore largely dependent on the interactions between particles.

We define N as the the number of interparticle interactions per unit volume, and so N_0 denotes the value of N at equilibrium (i.e., when zero shearing $\dot{\gamma} = 0$, suspension is at rest). It is important to note that there is a driving force for particles to return to their equilibrium state. Therefore, at an arbitrary point in time, the rate of formation of interaction points can be expressed in terms of the driving force to return to the equilibrium number of interaction points. In the following equation, k_1 is the rate constant for formation of interaction points:

$$\left. \frac{dN}{dt} \right|_{\text{formation}} = k_1 (N_0 - N)$$

With increased shear rates, we observe a reduction in the number of interaction points between particles since shearing breaks up particle clusters and disentangle (or straighten out) polymer chains. This may be expressed mathematically as a rate of loss of interactions between particles as shown below.

$$\text{rate of loss} = \frac{dN}{dt}\bigg|_{\text{loss}} = k_2 N \dot{\gamma}^m$$

The rate of loss of interactions due to shearing in the above rate equation is first order with respect to the instantaneous number of interactions per unit volume which is denoted N and m^{th} order with respect to the shear rate denoted $\dot{\gamma}$. The parameter k_2 represents the rate constant for loss of interaction points.

At an arbitrary point in time, we then have the combined effect of formation and loss of interactions, resulting in a net increase of particle interactions as follows:

$$\frac{dN}{dt} = \frac{dN}{dt}\bigg|_{\text{formation}} - \frac{dN}{dt}\bigg|_{\text{loss}} = k_1(N_0 - N) - k_2 N \dot{\gamma}^m$$

Under steady-state flow, all derivatives with respect to time would be zero. Therefore, we have the equation as a constraint shown below.

$$\frac{dN}{dt}\bigg|_{ss} = 0$$

$$k_1(N_0 - N) = k_2 N \dot{\gamma}^m$$

$$\frac{N_0 - N}{N} = \frac{k_2}{k_1}\dot{\gamma}^m$$

$$\frac{N}{N_0} = \frac{1}{1 + \frac{k}{k_1}\dot{\gamma}^m}$$

Now we can relate N which represents the volumetric number of interparticle interactions in the suspension, with apparent viscosity of the suspension, η_{app}. We note that the total viscosity of the suspension can be said to be a sum of two components: the viscosity of the base fluid $\eta_{\text{base fluid}}$ and the viscosity η_i contributed by interparticle interactions. In the Cross model, it is assumed that the base fluid is Newtonian and is present at all times.

$$\eta_{\text{app}} = \eta_{\text{base fluid}} + \eta_i$$

At this point we can also define some reference points for viscosity, whereby the apparent viscosity at zero shear and very high shear rates are η_0 and η_∞, respectively. We note here that at very high shear rates, viscosity of the suspension plateaus to a constant value η_∞ which is close to the viscosity of the base fluid. Therefore, $\eta_{\text{base fluid}} = \eta_\infty$ and we obtain the following.

$$\eta_{app} = \eta_\infty + \eta_i$$

$$\eta_i = \eta_{app} - \eta_\infty \tag{1}$$

We can now express η_i in terms of η_0 by noting that at zero shear, η_{i0} is related to total apparent viscosity η_0 as follows:

$$\eta_0 = \eta_\infty + \eta_{i0}$$

$$\eta_{i0} = \eta_0 - \eta_\infty \qquad (2)$$

Going back to our earlier expression, we note from the problem that we may assume a linear dependence between N and viscosity contribution by the particles η_i. Therefore, we obtain the following expressions.

$$\frac{N}{N_0} = \frac{\eta_i}{\eta_{i0}} = \frac{1}{1 + \frac{k_2}{k_1}\dot{\gamma}^m}$$

Finally, we can express viscosity contributions by particles (η_i and η_{i0}) in terms of total suspension viscosity by substituting (1) and (2) into the above expression. Finally, we have derived the Cross equation in the form as given.

$$\frac{\eta_i}{\eta_{i0}} = \frac{\eta_{app} - \eta_\infty}{\eta_0 - \eta_\infty} = \frac{1}{1 + \frac{k_2}{k_1}\dot{\gamma}^m}$$

$$\eta_{app} - \eta_\infty = \frac{\eta_0 - \eta_\infty}{1 + \alpha\dot{\gamma}^m}, \quad \alpha = \frac{k_2}{k_1}$$

(b) Suspensions of particles in liquids are ubiquitous in practical applications, including the pharmaceutical, food, cosmetics, and plastics industries. The rheology of particle suspensions is also studied to understand natural phenomena, such as volcanic lava and mudflows. The flow characteristics of suspensions depend on various factors including particle volume fraction, particle shape and size, interactions between particles, and the bulk flow pattern.

We can better understand the development of this solution by plotting a graph of viscosity against shear rate (or strain rate), for a non-Newtonian suspension modelled using the Cross model. The shape of the curve below shows shear-thinning, as viscosity decreases with increasing shear rate.

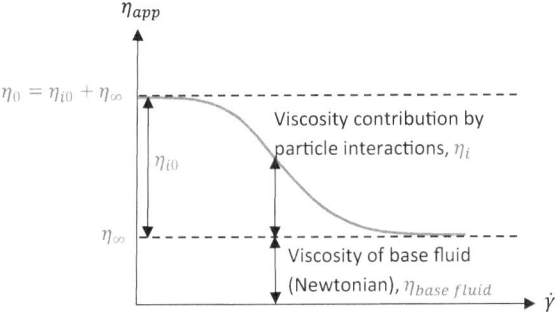

One explanation for shear-thinning in particle suspensions (e.g., non-colloidal suspensions) is based on the concept of *localized shear heating*. When there is shearing bulk flow in a suspension with a moderate to high volume fraction of particles, the particles approach close to one another causing localized squeezing flow, and very high local shear rates start to develop. This flow is "squeezed" as it occurs through the narrow gap between particles, and this leads to localized heating of the suspending fluid due to viscous dissipation. The localized heating reduces viscosity of liquid in the gap, which helps "smooth (or lubricate)" the path of particles past each other. As bulk shear rate increases, local shear rate increases, which results in increased viscous heating and a reduced viscosity (shear-thinning). As the viscosity of concentrated suspensions is typically dominated by viscous forces in the interparticle gaps, the localized shear heating mechanism explains the observed shear-thinning of total suspension viscosity.

It is however worth noting that it is assumed in our analysis in part a that the above-described localized heating does not cause a significant rise in bulk temperature of the suspension (i.e., constant temperature assumed).

Problem 27

Viscoelasticity of fluids can be determined experimentally using measuring apparatus called rheometers.

(a) **With suitable illustrations, briefly describe how the cone and plate rheometer and parallel plate rheometer may be used to determine shear stress and compare the two methods.**

(b) **Consider a cone and plate rheometer with a cone angle of 3° relative to the bottom plate. The radius of the circular cross section of the rheometer is 30 mm, and the following data were provided for the torque measured at varying rotation speeds of the upper cone.**

Rotation speed [revolutions/s]	0	0.057	0.19	0.26
Torque [Nm]	0	7.2×10^{-3}	18.9×10^{-3}	23.7×10^{-3}

(i) **Derive expressions for shear rate $\dot{\gamma}$ and shear stress τ.**

(ii) **Using the data provided, and assuming that the test fluid is well modelled by the power law equation shown below, determine viscosity η, the power law parameters (k and n), and the critical shear stress τ^* which marks a change in rheological behavior of the fluid. Present your findings in suitable plots.**

$$\tau = k\dot{\gamma}^n$$

Solution 27

Worked Solution

Parallel Plate Rheometer

(a) A parallel plate rheometer creates a space between two parallel surfaces through which a test fluid flows. One of the parallel surfaces (top disc in the example illustrated below) is made to rotate with an externally applied force, while the other surface is held stationary. Shear stress developed in the fluid may then be mathematically correlated to the magnitude of the applied force which is a measurable quantity. This force can be represented by torque, which may be thought of as the equivalent of force but used mainly for rotating systems.

In the diagram below, we have a parallel plate rheometer consisting of two parallel circular discs with radius R separated by a small gap of size H ($H \ll R$). The upper disc has a rotational speed (or angular velocity) of ω [radians per second] which causes shearing flow in the test fluid between the discs.

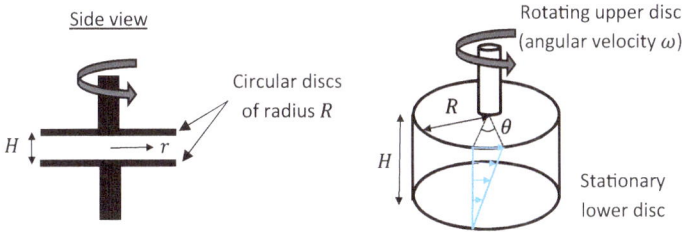

The flow profile of the test fluid is indicated by blue arrows shown above for a small angle of rotation θ [radians]. Using simple geometry, we can express strain γ and strain rate $\dot{\gamma}$ in terms of the displacement of fluid particles (refer to blue arrows pointing in the direction of flow) relative to the gap size H.

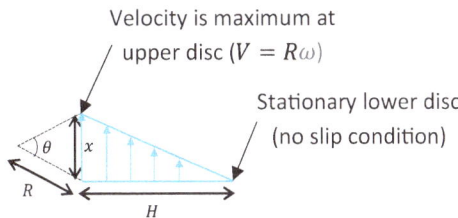

$$\gamma|_{\text{upper disc}} = \frac{x}{H} = \frac{R\theta}{H}, \quad \gamma|_{\text{lower disc}} = 0$$

$$\dot{\gamma}\big|_{\text{upper disc}} = \frac{R\omega}{H}, \quad \dot{\gamma}\big|_{\text{lower disc}} = 0$$

By definition, shear stress is shear force per unit area over which the force acts. For circular flow, we can express this area at an arbitrary distance r from the center of rotation as the area of a differential ring element.

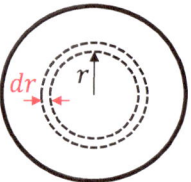

$$dA = 2\pi r\, dr$$

Torque T is a product of shear force and the perpendicular distance of this force from the center of rotation. Therefore, we can express torque dT at an arbitrary distance r from the center as follows.

$$dT = \tau A r = \tau\left(2\pi r^2\right) dr$$

In order to compute the value of the total measured torque T, we need to integrate the above equation over the entire disc, i.e., for all values of r from 0 to R.

$$T = \int_0^R \tau\left(2\pi r^2\right) dr = 2\pi \int_0^R \tau\left(r^2\right) dr$$

Note that for a parallel plate rheometer, both strain and strain rate are functions of r. Since shear stress $\tau = \eta\dot{\gamma}$ where η denotes viscosity, it follows that shear stress is also a function of r and so the term τ cannot be assumed constant in this integration. This may be mathematically more cumbersome than the cone and plate rheometer, which offers simpler mathematics by removing the dependence on r for strain and strain rate.

Cone and Plate Rheometer

A cone and plate rheometer differs from the parallel plate rheometer in that the top rotating disc is now a downward pointing cone whereby the tip of the cone just touches the center of the bottom disc as shown below.

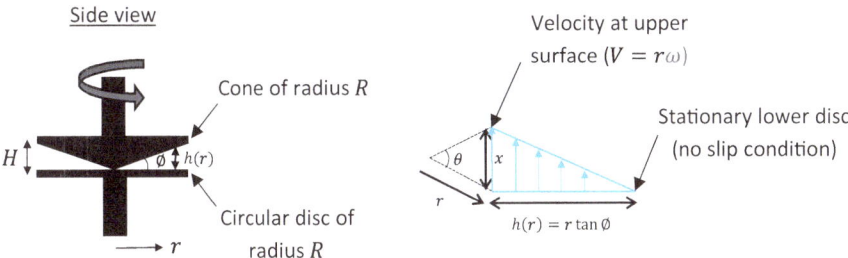

Side view

Cone of radius R

Circular disc of radius R

H Ø $h(r)$

r

Velocity at upper surface $(V = r\omega)$

Stationary lower disc (no slip condition)

θ x

r

$h(r) = r \tan \emptyset$

Even though this rheometer offers simpler mathematics, it may be more challenging to fabricate since it is difficult to achieve the tight fit (whereby the cone tip just touches the plate) between the cone and the bottom plate without damaging the apparatus.

Another difference between the cone and plate rheometer and the parallel plate rheometer lies in the presence of a cone angle $\emptyset$ which means the gap size is a function of radial distance denoted here as $h(r)$, instead of a constant value of h $(r) = H$. Therefore, the strain and strain rates for the upper and lower surfaces are adjusted as follows.

$$\gamma|_{\text{upper disc}} = \frac{x}{h} = \frac{R\theta}{R \tan \emptyset} = \frac{\theta}{\tan \emptyset}, \quad \gamma|_{\text{lower disc}} = 0$$

$$\dot{\gamma}|_{\text{upper disc}} = \frac{R\omega}{R \tan \emptyset} = \frac{\omega}{\tan \emptyset}, \quad \dot{\gamma}|_{\text{lower disc}} = 0$$

We notice that strain and strain rates are constant with respect to radial distance r. This makes the calculation of torque mathematically easier since shear stress is also independent of r and can be removed from the integral.

$$T = \int_0^R \tau\left(2\pi r^2\right)dr = 2\pi\tau\left[\frac{r^3}{3}\right]_0^R = \frac{2}{3}\pi\tau R^3$$

Since torque is easily measurable, we can relate torque to the shear stress τ with the above expression.

(b) (i) We can illustrate the setup as shown below to better visualize it.

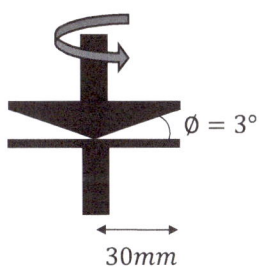

Ø $= 3°$

30mm

From our analysis in part a, we can express shear rate as follows for a cone and plate rheometer at an arbitrary radial distance r.

$$\dot{\gamma} = \frac{r\omega}{r\tan\varnothing} = \frac{\omega}{\tan\varnothing}$$

Let us denote rotation speed [revolutions per second] as f. We can relate angular velocity ω [radians per second] to f since one revolution goes through 2π radians; therefore, we can express shear rate as a function of f.

$$\dot{\gamma}\left[s^{-1}\right] = \frac{\omega}{\tan\varnothing} = \frac{2\pi f}{\tan\varnothing} = \left[\frac{2\pi}{\tan(3°)}\right]f \approx 120f \tag{1}$$

Also from our earlier analysis in part a, we have the expression for torque T below, which we can use to relate to shear stress τ.

$$T = \frac{2}{3}\pi\tau R^3$$

$$\tau[Pa] = \frac{3T}{2\pi R^3} = \frac{3T}{2\pi(30 \times 10^{-3})^3} = 1.77 \times 10^4 \, T \tag{2}$$

We are given the following data values for f and T. Using the expressions (1) and (2) above, we can derive values for $\dot{\gamma}$ and τ.

f[rev.s^{-1}]	0	0.057	0.19	0.26
T[Nm]	0	7.2×10^{-3}	18.9×10^{-3}	23.7×10^{-3}
$\dot{\gamma}$[s^{-1}]	0	$120(0.057) = 6.84$	22.8	31.2
τ[Pa]	0	$(1.77 \times 10^4)(7.2 \times 10^{-3}) = 127$	335	419

(b) (ii) For a power law fluid, there is Newtonian behavior at very low shear rates. This means that shear stress scales linearly with shear rate and the fluid exhibits constant viscosity η. We can compute an approximate value for viscosity by taking the data point with the lowest shear rate.

$$\eta = \frac{\tau}{\dot{\gamma}} = \frac{127}{6.84} = 18.6 \text{ Pas}$$

As shear rate increases past a critical value to intermediate to high values, the fluid will start to display a power law correlation between shear stress and shear rate as indicated by the model equation below where k and n are the power law parameters.

$$\tau = k\dot{\gamma}^n$$

Using the remaining two data points available for intermediate to high shear rates, we obtain two simultaneous equations that allow us to solve for the two unknowns (k and n).

$$335 = k(22.8)^n$$

$$\ln 335 = \ln k + n \ln 22.8 \tag{A}$$

$$419 = k(31.2)^n$$

$$\ln 419 = \ln k + n \ln 31.2 \tag{B}$$

Subtracting Eq. (A) from Eq. (B), we have

$$\ln \left(\frac{419}{335}\right) = n \ln \left(\frac{31.2}{22.8}\right)$$

$$n = 0.71$$

Note that this fluid is shear-thinning, since $n < 1$, which means that viscosity reduces with increasing shear rate in the power law region. Substituting this value of n back into one of the equations (in this case we arbitrarily chose Eq. (B)), we obtain the value of k.

$$\ln 419 = \ln k + 0.71 \ln 31.2$$

$$k = 36$$

The power law model equation is therefore

$$\tau = 36 \dot{\gamma}^{0.71}$$

We can now determine the critical shear stress τ^* as it must hold true that the shear stress function is a continuous function for it to physically make sense. This means that at the critical shear rate $\dot{\gamma}^*$, the shear stress value computed from the Newtonian model equation must be equivalent to the shear stress value computed from the power law model equation, and this value is simply τ^*.

$$\tau^* = \eta \dot{\gamma}^* = k \dot{\gamma}^{*n}$$

$$18.6 \dot{\gamma}^* = 36 \dot{\gamma}^{*0.71}$$

$$\dot{\gamma}^* = \left(\frac{36}{18.6}\right)^{\frac{1}{0.29}} = 9.7 \ s^{-1}$$

$$\tau^* = 18.6(9.7) = 180 \ Pa$$

We can present the quantities found above in plots of shear stress τ and viscosity (more precisely apparent viscosity η_{app} which is also the experimentally observed viscosity).

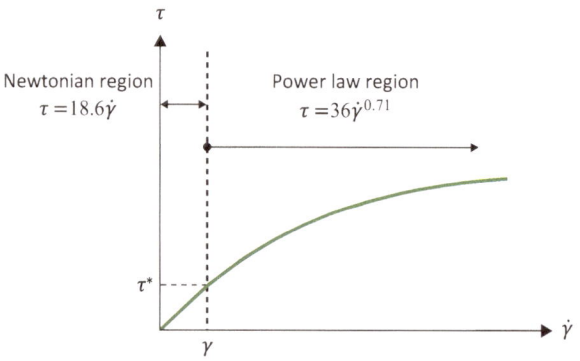

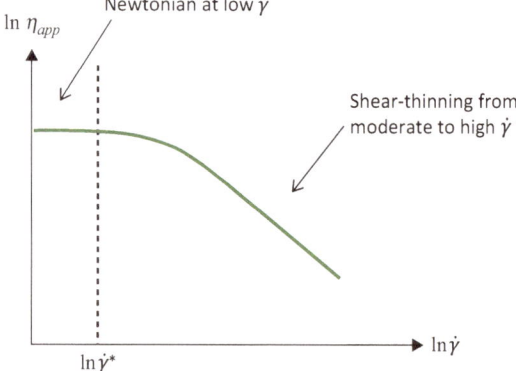

Problem 28

In a controlled strain experiment for a viscoelastic fluid, an oscillatory strain was applied to the fluid, and the resultant shear stress developed in the fluid was measured. Given that the applied strain γ may be described by the cosine function below, where ω denotes angular frequency and $\varnothing$ denotes phase angle of the stress response relative to applied strain, (you may assume that $\varnothing$ is a small angle).

$$\gamma = \gamma_0 \cos\omega t$$

(a) **Show that the stress response may be expressed as follows whereby G' and G'' are moduli analogous to the proportionality constant k in Hooke's law in the equation governing the extension of an elastic spring (applied force F is related to extension length x according to $F = kx$).**

$$\tau = G'\gamma_0\cos\omega t - G''\gamma_0\sin\omega t$$

(b) **Given that $\omega = 1$ radian per second, explain how the stress response function relates to the strain and strain rate functions using a suitable plot.**

(c) **By considering the complex modulus G^*, derive the following expressions and comment on the physical significance of the real and imaginary parts of G^*.**

$$G' = \frac{\tau_0}{\gamma_0}\cos\varnothing$$

$$G'' = \frac{\tau_0}{\gamma_0}\sin\varnothing$$

Solution 28

Worked Solution

(a) We are given the sinusoidal function for applied strain (or deformation) which we can use to deduce strain rate.

$$\gamma = \gamma_0 \cos \omega t$$

$$\frac{\gamma}{\gamma_0} = \cos \omega t$$

Strain rate $\dot{\gamma}$ is simply the derivative of strain with respect to time as shown below.

$$\dot{\gamma} = \frac{d\gamma}{dt} = -\gamma_0\omega \sin \omega t$$

$$\dot{\gamma} = -\dot{\gamma}_0 \sin \omega t, \quad \dot{\gamma}_0 = \gamma_0\omega$$

$$\frac{\dot{\gamma}}{\dot{\gamma}_0} = - \sin \omega t$$

Let us pause here to examine the significance of the strain function γ and the strain rate function $\dot{\gamma}$. We can compare the time dependence of both functions by plotting the normalized curves of $\frac{\gamma}{\gamma_0}$ and $\frac{\dot{\gamma}}{\gamma_0}$ against time t. [For simplicity, let us assume that $\omega = 1$ rad.s^{-1} so that our plot against t is equivalent to that against ωt.]

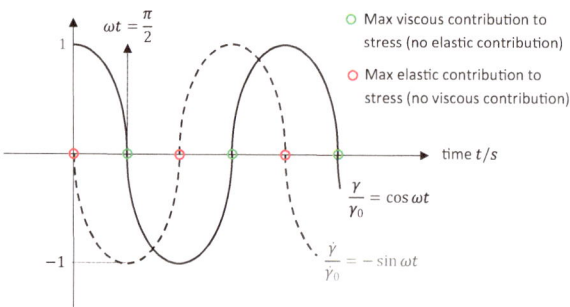

Viscous and Elastic Components

Recall that we can analyze viscoelastic fluids by breaking them down into a fully elastic component and a fully viscous component. The elastic component can be modelled using Hooke's law whereby, like a spring, shear stress is directly proportional to strain where the proportionality constant is an elastic constant (denoted g here).

$$\tau = g\gamma$$

The viscous component can be modelled after a linear Newtonian fluid exhibiting constant viscosity. In this instance, shear stress is directly proportional to strain rate whereby the proportionality constant is viscosity (denoted η here).

$$\tau = \eta\dot{\gamma}$$

Therefore, the viscous contribution to stress scales with strain rate ($\tau = \eta\dot{\gamma}$, $\tau \sim \dot{\gamma}$), while the elastic component to stress scales with strain ($\tau = g\gamma$, $\tau \sim \gamma$). The total stress response of the viscoelastic fluid will be the sum of these two contributions.

Key Features of the Strain and Strain Rate Curves

By observing the earlier plot, we notice the following:

1. When $\frac{\gamma}{\gamma_0}$ is at maximum amplitude $\rightarrow$ maximum elastic component to stress (points circled in red)

- Occurs when $\frac{\dot{\gamma}}{\dot{\gamma}_0}$ is zero, which is when the viscous component to stress is minimum (zero).

2. When $\frac{\gamma}{\gamma_0}$ is zero → minimum (zero) elastic component to stress (points circled in green)

- Occurs when $\frac{\dot{\gamma}}{\dot{\gamma}_0}$ is maximum, which is when the viscous component to stress is maximum.

3. All other times other than 1 and 2 above, i.e., periods between the instances circled in red and green

- Non-zero contributions from both the elastic and viscous components of stress to the overall stress response.

We are also told that the stress response has a small phase angle of $\varnothing$ relative to the strain profile; therefore, we can express shear stress as follows, which means that the stress curve is displaced from the strain curve by a phase difference of $\varnothing$ radians.

$$\tau = \tau_0 \cos{(\omega t + \varnothing)}$$

The expression above can be expanded using the trigonometric addition formula. This step helps us to separate the argument (i.e., $\omega t + \varnothing$) of the cosine function into two physically meaningful parts. The first term is in phase with strain since it similarly varies with time according to "$\cos \omega t$," and the second term is in phase with strain rate since it similarly varies with time according to "$- \sin \omega t$."

$$\tau = \tau_0 \cos \omega t \cos \varnothing + (-\tau_0 \sin \omega t \sin \varnothing)$$

We notice that this is in line with our earlier analysis where we observed the viscous contribution to stress and elastic contribution to stress from the shapes of the plots for strain and strain rate. We can slightly rearrange the equation to separate the time-dependent terms from the time-independent terms. Specifically, the time-independent portion relates to the maximum amplitude as highlighted in yellow below, and the time-dependent terms are highlighted in gray below.

$$\tau(t) = (\tau_0 \cos \varnothing)(\cos \omega t) + (\tau_0 \sin \varnothing)(- \sin \omega t)$$

$(\tau_0 \cos \varnothing)$ represents the maximum amplitude (or magnitude) of stress contribution from elasticity, while $(\tau_0 \sin \varnothing)$ denotes the maximum amplitude (or magnitude) of stress contribution from viscosity. Mathematically, we can further express these maximum amplitude terms in terms of moduli G' and G'' whereby the moduli are analogous to the proportionality constant in Hooke's law (a proxy "elastic constant") that relates stress to strain (recall $\tau \sim \gamma$).

$$\tau_0{}^{\text{elastic}} = \tau_0 \cos \varnothing = G'\gamma_0$$

$$\tau_0{}^{\text{viscous}} = \tau_0 \sin \varnothing = G''\gamma_0$$

It makes sense therefore that we define G' as the elastic (or storage) modulus since it contributes to the elastic component to stress and we define G'' as the viscous (or loss) modulus since it contributes to the viscous component to stress.

$$G' = \frac{\tau_0 \cos \varnothing}{\gamma_0}, \quad G'' = \frac{\tau_0 \sin \varnothing}{\gamma_0}$$

Going back to our earlier expression, and putting all our results together, we can express the overall stress response with time in terms of the two moduli as shown below.

$$\tau = (\tau_0 \cos \varnothing)(\cos \omega t) + (\tau_0 \sin \varnothing)(-\sin \omega t)$$

$$\tau = \tau_0{}^{\text{elastic}}(\cos \omega t) + \tau_0{}^{\text{viscous}}(-\sin \omega t)$$

$$\tau = G'\gamma_0 \cos \omega t + G''\gamma_0(-\sin \omega t)$$

(b) We will be able to visualize the relationships between strain, strain rate, and stress response for an oscillatory deformation by plotting the normalized graphs of each quantity, i.e., $\frac{\gamma}{\gamma_0}, \frac{\dot{\gamma}}{\gamma_0}$ and $\frac{\tau}{\tau_0}$.

We have earlier sketched the graphs of $\frac{\gamma}{\gamma_0}$ and $\frac{\dot{\gamma}}{\gamma_0}$ in part a. Extending from there, we note that the normalized stress response function is in fact a cosine curve shifted by a small angle $\varnothing$ to the left, as shown in brown below.

$$\frac{\tau}{\tau_0} = \cos(\omega t + \varnothing), \quad 0 < \varnothing < \frac{\pi}{2}$$

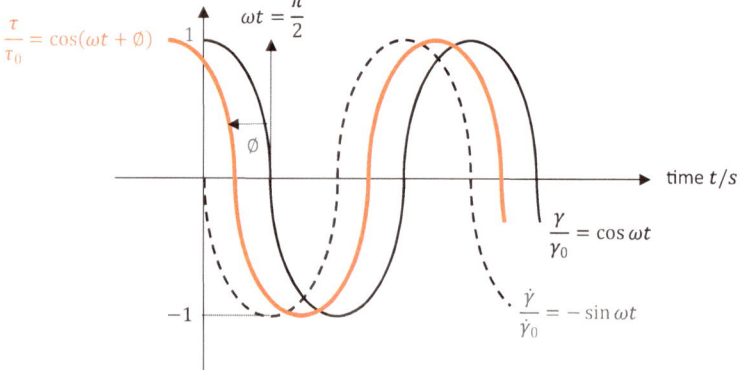

Note that the phase angle is expected to lie within the range $0 < \varnothing < \frac{\pi}{2}$ such that the stress curve lies in between the two limits set by the strain and strain rate curves.

Since strain and strain rates are both continuous trigonometric functions with time, and since strain relates to elasticity while strain rate relates to viscosity, both these contributions are active (although at varying degrees at different points in time) to overall stress as time progresses. The viscous and elastic contributions are not expected to stagnate (e.g., zero at all times or remain maximum at all times) with time. This gives rise to the similarly oscillatory stress response. This also explains the limits that phase angle $\varnothing$ takes.

If $\varnothing = 0$ (elastic limit), the fluid is akin to being fully elastic at all times $t > 0$, which is not physically sensible for a viscoelastic fluid that experiences both viscous and elastic contributions to stress as time progresses.

$$\tau = \tau_0 \cos \omega t, \quad \varnothing = 0$$

Similarly, if $\varnothing = \frac{\pi}{2}$ (viscous limit), the fluid is akin to being fully viscous at all times $t > 0$, which is not physically sensible for a viscoelastic fluid that experiences both viscous and elastic contributions to stress as time progresses.

$$\tau = \tau_0(-\sin \omega t), \quad \varnothing = \frac{\pi}{2}$$

(c) Recall that a complex exponential describes a complex number (denoted z here) that comprises both a real part and an imaginary part which can be written as trigonometric expressions (polar form) in terms of an angle of rotation θ from the horizontal real axis as illustrated below. The diagram below is also often referred to as the Argand diagram.

$$z = r(\cos \theta + i \sin \theta) = re^{i\theta}$$

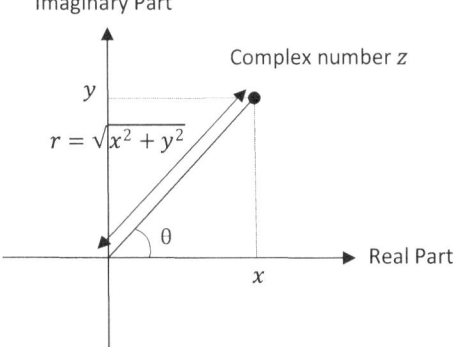

We can express strain in its complex form γ^* as follows, whereby strain (being a physically meaningful quantity) can then be equated to the real part of its complex form.

$$\gamma^* = re^{i\theta} = r(\cos\theta + i\sin\theta)$$

$$\gamma = \text{Re}\,(\gamma^*) = r\cos\theta$$

Comparing the above expression with the given strain function, it follows that the complex form of strain can be expressed in the complex exponential form as shown.

$$\gamma = \gamma_0\cos\omega t \rightarrow r = \gamma_0, \theta = \omega t$$

$$\gamma^* = \gamma_0 e^{i\omega t}$$

Mathematically, exponential terms are relatively easy to integrate and differentiate (with respect to time in this case). We can differentiate the complex form of strain γ^* to obtain the complex form of strain rate $\dot{\gamma}^*$ as shown below.

$$\dot{\gamma}^* = \frac{d}{dt}(\gamma^*) = \frac{d}{dt}\left(\gamma_0 e^{i\omega t}\right) = \gamma_0 i\omega e^{i\omega t} = \gamma_0 i\omega(\cos\omega t + i\sin\omega t)$$

Taking the real part of the complex form of strain rate would give us the physically meaningful strain rate function, which is the same as that found in the earlier part of this problem. Note that by definition, $i^2 = -1$.

$$\dot{\gamma} = \text{Re}\,[\dot{\gamma}^*] = \text{Re}\left[\gamma_0 i\omega\cos\omega t + \gamma_0\omega(i^2)\sin\omega t\right] = \text{Re}\left[\gamma_0 i\omega\cos\omega t - \gamma_0\omega\sin\omega t\right]$$

$$\dot{\gamma} = -\gamma_0\omega\sin\omega t$$

Similarly for the stress response, we can express stress τ in its complex form by considering the complex forms for the modulus and strain terms in the correlation below. We can see the physical significance of the real and imaginary parts of the complex modulus G^* in that the real part is in fact the storage (or elastic) modulus G' and the imaginary part corresponds to the loss (or viscous) modulus G''.

$$\tau^* = G^*\gamma^*$$

$$G^* = G' + iG''$$

Separately, by definition of the complex exponential, we can express stress in its complex exponential form.

$$\tau^* = \tau_0 e^{i(\omega t + \varnothing)}$$

$$\frac{\tau^*}{\gamma^*} = \frac{\tau_0 e^{i(\omega t + \varnothing)}}{\gamma_0 e^{i\omega t}} = G' + iG''$$

$$\frac{\tau_0 e^{i\varnothing}}{\gamma_0} = G' + iG''$$

$$\frac{\tau_0}{\gamma_0}(\cos\varnothing + i\sin\varnothing) = G' + iG''$$

By comparing both sides of the equation and equating the corresponding real and imaginary parts, we obtain the same expressions for the moduli as before.

$$G' = \frac{\tau_0}{\gamma_0}\cos\varnothing, \quad G'' = \frac{\tau_0}{\gamma_0}\sin\varnothing$$

Problem 29

In another controlled strain experiment for a test fluid with viscoelastic properties, a sinusoidal deformation was applied according to the function below, and the resultant stress response was found to have a phase difference of $\varnothing$ radians relative to the input strain where $\varnothing$ is a small angle. In the expression below, ω denotes angular frequency with the value of 1 radian per second.

$$\gamma = \gamma_0 sin\omega t$$

(a) **Show that the same expressions for the real and imaginary parts of the complex modulus G^* as in Problem 28 (i.e., cosine function for strain) may be derived for this sine function for input strain as shown below.**

$$G' = \frac{\tau_0}{\gamma_0}cos\varnothing, \quad G'' = \frac{\tau_0}{\gamma_0}sin\varnothing$$

(b) **Comment on factors that would affect the expressions in part a.**

Solution 29

Worked Solution

(a) In this problem, we will be able to demonstrate that the expressions for the storage (or elastic) modulus denoted G' and loss (or viscous) modulus G'' are the same for a cosine or sine input strain function, as they are both oscillatory deformations.

This time we have a sine function for applied strain, we can again determine the strain rate function which is the derivative of strain with respect to time. γ_0 denotes the maximum strain amplitude.

$$\gamma = \gamma_0 \sin \omega t$$

$$\dot{\gamma} = \frac{d\gamma}{dt} = \gamma_0 \omega \cos \omega t$$

$$\dot{\gamma} = \dot{\gamma}_0 \cos \omega t, \quad \dot{\gamma}_0 = \gamma_0 \omega$$

$$\frac{\dot{\gamma}}{\dot{\gamma}_0} = \cos \omega t$$

The output stress response has a phase difference of angle $\varnothing$ relative to strain at all times; therefore, we express shear stress as follows.

$$\tau = \tau_0 \sin (\omega t + \varnothing)$$

As before, we expand the expression into two distinct parts, one representing the elastic contribution that is in phase with strain and the other viscous contribution that is in phase with strain rate.

Overall stress $\tau =$ component in phase with strain + component in phase with strain rate

$$\tau(t) = (\tau_0 \cos \varnothing)(\sin \omega t) + (\tau_0 \sin \varnothing)(\cos \omega t)$$

The parts highlighted in gray above are time-dependent and indicate terms that are in phase with strain and strain rate, respectively. The parts highlighted in yellow are time-independent and tell us the maximum amplitudes for stress for each contribution. Specifically, $(\tau_0 \cos \varnothing)$ represents the maximum amplitude (or magnitude) of stress contribution from elasticity, while $(\tau_0 \sin \varnothing)$ denotes the maximum amplitude (or magnitude) of stress contribution from viscosity.

Like before, we can express these maximum amplitude terms in terms of moduli G' and G'' where the moduli are analogous to the proportionality constant in Hooke's law (a proxy "elastic constant") that relates stress to strain (recall $\tau \sim \gamma$).

$$\tau_0{}^{elastic} = \tau_0 \cos \varnothing = G'\gamma_0 \rightarrow G' = \frac{\tau_0 \cos \varnothing}{\gamma_0}$$

$$\tau_0{}^{viscous} = \tau_0 \sin \varnothing = G''\gamma_0 \rightarrow G'' = \frac{\tau_0 \sin \varnothing}{\gamma_0}$$

(b) We note that by changing the strain input from a cosine (in previous Problem 28) to sine function had no impact on the expressions for the real and imaginary parts of the complex modulus.

For a given strain input function, the maximum strain amplitude γ_0 is defined. The stress response is an output that can be obtained experimentally from rheometric measurements. Hence, we can observe and record values for τ_0 and $\emptyset$ from a rheometer for the given strain input.

The storage and loss moduli in the expression for stress response contain time-independent quantities. The change in input strain function from cosine to sine acted to change the shear stress response from cosine to sine, i.e., the time-dependent portion was adjusted to account for this change.

However, the moduli remained unchanged given the same phase difference of $\emptyset$ for both the first case of a cosine function (in previous Problem 28) and the second case of a sine function (current Problem 29). The maximum amplitude contributions of both the elastic and viscous parts of the fluid remained the same.

Factors Affecting Storage and Loss Moduli

Some factors that might alter the values of the storage and loss moduli include temperature, strain intensity (related to maximum strain amplitude for input strain), and angular frequency of oscillatory deformation.

Temperature

Experimental studies done on viscoelastic fluids such as the PR 520 (an epoxy resin system) have shown that over a defined range, the storage (or elastic) modulus decreased with increasing temperature. Conversely, the loss (or viscous) modulus was found to increase with increasing temperature. In other words, a higher temperature led to reduced strength and elasticity and a greater viscous contribution to stress.

Strain Intensity

The modulus is also dependent on the intensity of the strain input which relates to the degree of deformation. As strain intensity increased, the storage modulus decreased, while the reverse is true for the loss modulus. In other words, the material becomes less stiff or elastic as the degree of deformation was greater.

Angular Frequency

As for the angular frequency of oscillations of input strain, the greater the frequency, the higher the storage modulus, while the reverse is true for the loss modulus. This means that as the strain value varied more rapidly, the substance tended toward more elastic behavior, becoming stiffer or in other words "putting up a greater resistance" against the externally applied deformation.

Problem 30

Two fluid samples were tested using a rheometer to identify which is mayonnaise and which is mineral oil. A long and narrow cylindrical shaft with a radius of 15 mm is lowered vertically into the test fluid up to a depth of 150 mm. The shaft is made to rotate at three revolutions per second, and the measured torque on the shaft is 0.4 Nm.

(a) Given that both fluids exhibit linear behavior, explain how the fluid samples can be differentiated and identified, citing appropriate model equations and expected results for each of the test fluids.

   If the test fluid was mineral oil,

(b) Show that the angular velocity of fluid motion driven by the rotating shaft may be expressed as follows with respect to radial distance r:

$$\omega(r) = 6\pi \frac{r_1{}^2}{r^2}$$

(c) Determine the shear rate at the boundary between the shaft and fluid and comment on the result.

(d) Determine the viscosity of the fluid.

   If the test fluid was mayonnaise,

(e) Show that the point where the fluid transitions from being stationary to having non-zero velocity can be expressed as follows. Comment on the physical significance of K in the expression.

$$r^* = \sqrt{\frac{4}{3\pi K}}$$

(f) Derive the angular velocity profile of the fluid with respect to radial distance.

(g) **Given that the value of K for the expression in part e is $K = 60$, determine the viscosity coefficient and shear rate at the boundary between shaft and fluid.**

(h) **Determine the apparent viscosity of the fluid at the boundary between shaft and fluid for the applied shear rate.**

(i) **Briefly discuss any downsides and possible improvements to the rheometric experiment described in this problem.**

Solution 30

Worked Solution

(a) Rheometric experiments can be used to characterize sample fluids based on their observed flow behavior. This problem requires a distinction between mayonnaise and mineral oil, and one key differentiating feature to note is that the former is a Bingham fluid, while the latter is a Newtonian fluid. [Note that Bingham fluids may be Newtonian or non-Newtonian (shear-thinning or shear-thickening). In this case, we assume a Newtonian Bingham fluid since the problem states that both fluids exhibit linearity.]

Most fluids used in the food industry are Bingham fluids. A key feature of these fluids is that they have a yield stress which we have denoted its magnitude as τ_y. [Note that there is a negative sign to account for the fact that shear stress in the fluid acts to oppose the applied shear stress on the fluid τ.] The fluid behaves like a solid when the applied shear stress τ is smaller than yield stress τ_y and behaves like a liquid when stress exceeds yield stress. Some examples of such fluids include mayonnaise, chocolate, and mustard. The characteristic equation describing such Bingham fluids is shown below, where η is the viscosity coefficient to shear rate $\dot{\gamma}$. [We have assumed here in the constitutive equation that the fluid is a Newtonian Bingham fluid for simplification reasons although many such real fluids are Bingham shear-thinning.]

$$\tau = -\tau_y + \eta\dot{\gamma}, \quad |\tau| > |\tau_y|$$

When $|\tau| < |\tau_y|$, $\dot{\gamma} = 0$, and the fluid behaves like a solid. When $|\tau| > |\tau_y|$, the fluid starts to behave like a Newtonian fluid where shear stress scales linearly with shear rate.

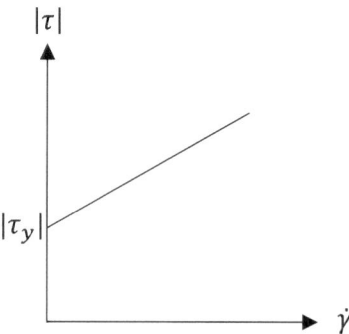

As for Newtonian fluids (non-Bingham), some common examples include water and mineral oil (in this problem). Their rheological behavior is relatively simple, as shear stress scales linearly with shear rate with a proportionality constant equivalent to viscosity η.

$$\tau = \eta\dot{\gamma}$$

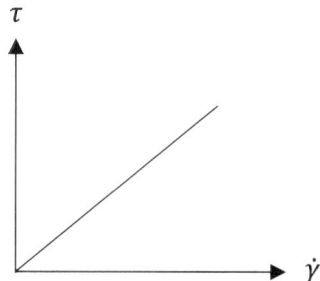

(b) Let us illustrate our experimental setup below to better visualize it.

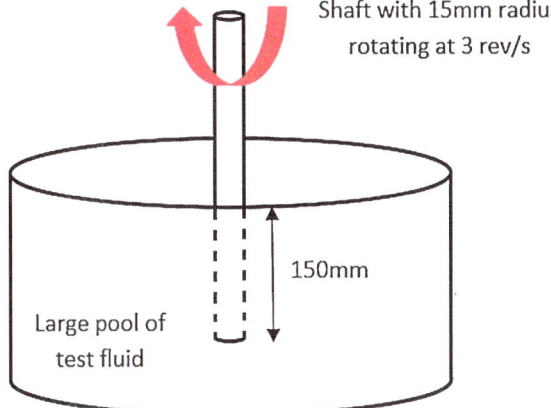

Shaft with 15mm radius
rotating at 3 rev/s

150mm

Large pool of
test fluid

If the fluid was mineral oil, it would follow the stress function as expressed below.

$$\tau = \eta\dot{\gamma}$$

In order to find the angular velocity profile, we can start by examining the relationship of its derivative to strain rate and consider boundary conditions to integrate the differential equation. By definition, we can define strain rate for the fluid motion in our problem as follows.

$$\dot{\gamma} = r\frac{d\omega}{dr}$$

$$\tau = \eta r\frac{d\omega}{dr}$$

The measured torque T is the product of tangential force of rotation and its perpendicular distance to the center of rotation. Forces balance in a steady system; hence, the applied force of rotation just overcomes the shear stress in the fluid (i.e., shear force equal to applied rotational force). The force of rotation is simply the product of shear stress and surface area over which the stress acts. We can therefore write the force balance equation as follows where L denotes the depth of shaft immersed in the fluid.

$$\text{Net force} = 0 \rightarrow T + r(\tau)(2\pi r L) = 0$$

$$\tau = -\frac{T}{2\pi r^2 L}$$

We can now equate the two expressions for stress τ as follows.

$$\eta r \frac{d\omega}{dr} = -\frac{T}{2\pi r^2 L}$$

$$d\omega = -\frac{T}{2\pi L \eta}\left(\frac{1}{r^3}\right)dr$$

We are required to find the angular velocity profile $\omega(r)$, which can be obtained by integrating the above differential equation with boundary conditions. We note that when radial distance from the center of rotation is equivalent to radius of the rotating shaft (denoted $r_1 = 15$ mm here), the frequency of rotations $f = 3$.

$$\omega|_{r=r_1} = 2\pi f = 2\pi(3) = 6\pi$$

We can integrate the earlier differential equation from this boundary condition to an arbitrary distance r and angular velocity ω.

$$\int_{\omega|_{r=r_1}}^{\omega} d\omega = -\frac{T}{2\pi L \eta}\int_{r_1}^{r}\frac{1}{r^3}dr$$

$$\omega - \omega|_{r=r_1} = -\frac{T}{2\pi L \eta}\left[-\frac{1}{2r^2}\right]_{r_1}^{r} = -\frac{T}{2\pi L \eta}\left[-\frac{1}{2r^2} - \left(-\frac{1}{2r_1^2}\right)\right]$$

$$\omega - 6\pi = -\frac{T}{4\pi L \eta}\left[\frac{1}{r_1^2} - \frac{1}{r^2}\right]$$

Another boundary condition is at infinitely large distance away from the shaft, the fluid is stationary. We can substitute this into the expression above,

$$\omega|_{r=\infty} = 0$$

$$0 - 6\pi = -\frac{T}{4\pi L \eta}\left[\frac{1}{r_1^2} - 0\right]$$

$$\frac{T}{4\pi L \eta} = 6\pi r_1^2$$

We can use our result above to simplify our angular velocity profile expression to obtain the given expression.

$$\omega - 6\pi = -\frac{T}{4\pi L \eta}\left[\frac{1}{r_1^2} - \frac{1}{r^2}\right] = -6\pi r_1^2\left[\frac{1}{r_1^2} - \frac{1}{r^2}\right] = 6\pi\left[\frac{r_1^2}{r^2} - 1\right]$$

$$\omega = 6\pi\frac{r_1^2}{r^2}$$

(c) The shear rate at the boundary between the shaft and the fluid can be determined by recalling the shear rate equation and establishing the boundary condition $r = r_1$

$$\dot{\gamma} = r\frac{d\omega}{dr}$$

$$\frac{d\omega}{dr} = \frac{d}{dr}\left(6\pi\frac{r_1^{\,2}}{r^2}\right) = -12\pi\frac{r_1^{\,2}}{r^3}$$

$$\dot{\gamma} = r\left(-12\pi\frac{r_1^{\,2}}{r^3}\right) = -12\pi\frac{r_1^{\,2}}{r^2}$$

$$\dot{\gamma}|_{r=r_1} = -12\pi = -37.7s^{-1}$$

We notice that the shear rate at the boundary of the shaft and fluid is independent of the shaft radius r_1. The magnitude of shear rate is $\left|\dot{\gamma}|_{r=r_1}\right| = 37.7\ s^{-1}$.

(d) The measured torque $T = 0.4$ Nm; we can find viscosity η using our expression from part b earlier. $L = 150$ mm $= 0.15$ m and $r_1 = 15$ mm $= 0.015$ m.

$$\frac{T}{4\pi L\eta} = 6\pi r_1^{\,2}$$

$$\eta = \frac{0.4}{4\pi(0.15)\left(6\pi(0.015)^2\right)} = 50\ \text{Pa.s}$$

(e) If the fluid was mayonnaise, it would exhibit Bingham properties. This means that the model equation will be as follows as explained in part a. [Note that for simplification reasons, we assume here that mayonnaise is a Bingham Newtonian fluid. However, note that mayonnaise displays shear-thinning properties at higher shear rates and is more accurately a Bingham shear-thinning (non-Newtonian) fluid.]

$$\tau = -\tau_y + \eta\gamma;\ \ |\tau| > |\tau_y|$$

The measured torque $T = 0.4$ Nm; therefore, we can construct a similar force balance as before for the case of mineral oil, but this time, we will use the stress equation for a Bingham fluid instead.

$$\text{Net force} = 0 \rightarrow T + r(\tau)(2\pi r L) = 0$$

$$\tau = -\frac{T}{2\pi r^2 L}$$

At the point ($r = r^*$) where the fluid just transitions from being stationary to a state of motion is when the yield stress is reached. Therefore, $|\tau| = |\tau_y|$ when $r = r^*$, and substituting the magnitude of measured torque $T = 0.4$ Nm and $L = 0.15$ m, we arrive at the given expression.

$$|\tau| = |\tau_y| = \frac{T}{2\pi (r^*)^2 L}$$

$$r^* = \sqrt{\frac{0.4}{2\pi L |\tau_y|}} = \sqrt{\frac{0.4}{2\pi (0.15) |\tau_y|}} = \sqrt{\frac{4}{3\pi |\tau_y|}} = \sqrt{\frac{4}{3\pi K}}$$

The physical significance of the parameter "K" in the given expression is its representation of the magnitude of yield stress of mayonnaise.

(f) In order to determine the angular velocity profile for mayonnaise, we need to consider the region where $|\tau| > |\tau_y|$ since that is when there is fluid flow.

$$\tau = -\tau_y + \eta \dot{\gamma} = -\frac{T}{2\pi r^2 L}$$

$$\dot{\gamma} = r \frac{d\omega}{dr}$$

$$-\tau_y + \eta r \frac{d\omega}{dr} = -\frac{T}{2\pi r^2 L}$$

$$\int_{\omega|_{r=r_1}}^{\omega} d\omega = \int_{r_1}^{r} \left(\frac{\tau_y}{\eta r} - \frac{T}{2\pi L \eta r^3} \right) dr$$

$$\omega - \omega|_{r=r_1} = \frac{\tau_y}{\eta} \ln \left(\frac{r}{r_1} \right) + \frac{T}{4\pi L \eta} \left(\frac{1}{r^2} - \frac{1}{r_1^2} \right)$$

We know that $\omega|_{r=r_1}$ has the value of 6π from part a; therefore,

$$\omega = \frac{\tau_y}{\eta} \ln \left(\frac{r}{r_1} \right) + \frac{T}{4\pi L \eta} \left(\frac{1}{r^2} - \frac{1}{r_1^2} \right) + 6\pi \tag{1}$$

We now note the boundary condition that is characteristic for the Bingham fluid, whereby at the point of yield stress, $r = r^*$, and $\omega = 0$; therefore, we can also express $\omega|_{r=r_1}$ in terms of the yield stress position r^*.

$$\omega\big|_{r=r^*} = 0$$

$$0 - \omega\big|_{r=r_1} = 0 - 6\pi = \frac{\tau_y}{\eta} \ln\left(\frac{r^*}{r_1}\right) + \frac{T}{4\pi L\eta}\left(\frac{1}{(r^*)^2} - \frac{1}{r_1{}^2}\right)$$

$$\omega\big|_{r=r_1} = 6\pi = \frac{\tau_y}{\eta} \ln\left(\frac{r_1}{r^*}\right) + \frac{T}{4\pi L\eta}\left(\frac{1}{r_1{}^2} - \frac{1}{(r^*)^2}\right)$$

We can now substitute the above expression for 6π back into Eq. (1) to express angular velocity as a function of r as follows, where r_1 falls out of the expression and angular velocity is this time independent of the radius of the shaft, but is instead dependent on the radial distance at which yield stress is reached r^*.

$$\omega = \frac{\tau_y}{\eta} \ln\left(\frac{r}{r_1}\right) + \frac{T}{4\pi L\eta}\left(\frac{1}{r^2} - \frac{1}{r_1{}^2}\right) + \frac{\tau_y}{\eta} \ln\left(\frac{r_1}{r^*}\right) + \frac{T}{4\pi L\eta}\left(\frac{1}{r_1{}^2} - \frac{1}{(r^*)^2}\right)$$

$$\omega = \frac{\tau_y}{\eta} \ln\left(\frac{r}{r^*}\right) + \frac{T}{4\pi L\eta}\left(\frac{1}{r^2} - \frac{1}{(r^*)^2}\right)$$

(g) We have found from the earlier part e that

$$|\tau_y| = K = 60$$

$$r^* = \sqrt{\frac{4}{3\pi K}} = 0.084 \text{ m}$$

We can now determine the viscosity coefficient η, by establishing the boundary condition of when $r = r_1$ as we know $r_1 = 0.015$ m, $\omega\big|_{r=r_1} = 6\pi$, $T = 0.4$ Nm, and $L = 0.15$ m.

$$\eta = \frac{\tau_y}{\omega\big|_{r=r_1}} \ln\left(\frac{r_1}{r^*}\right) + \frac{T}{4\pi L\omega\big|_{r=r_1}}\left(\frac{1}{r_1{}^2} - \frac{1}{(r^*)^2}\right)$$

$$\eta = \frac{60}{6\pi} \ln\left(\frac{0.015}{0.084}\right) + \frac{0.4}{4\pi(0.15)6\pi}\left(\frac{1}{0.015^2} - \frac{1}{0.084^2}\right) = 43 \text{ Pa.s}$$

The shear rate at the boundary between the rotating shaft and the fluid is simply the condition for $\dot{\gamma}$ when $r = r_1$. Revisiting our force balance from earlier, and substituting values accordingly, we have the following:

$$\tau|_{r=r_1} = -\tau_y + \eta\dot\gamma|_{r=r_1} = -\frac{T}{2\pi r_1^2 L}$$

$$\dot\gamma|_{r=r_1} = \frac{\tau_y}{\eta} - \frac{T}{2\pi r_1^2 L\eta} = \frac{60}{43} - \frac{0.4}{2\pi 0.015^2(0.15)43} = -42.5 \text{ s}^{-1}$$

The magnitude for shear rate is found to be 42.5 s^{-1}.

(h) The apparent viscosity η_{app} refers to the "proxy" (or equivalent) viscosity if we assume a Newtonian fluid equation is used to model this Bingham fluid. But this viscosity is only "proxy" or apparent because it is in fact not a Newtonian fluid that obeys the equation below and is instead a Bingham fluid. Establishing the condition when $r = r_1$, we have

$$\tau|_{r=r_1} = \eta_{app}\dot\gamma|_{r=r_1}$$

$$-\frac{T}{2\pi r_1^2 L} = \eta_{app}\dot\gamma|_{r=r_1}$$

$$-\frac{0.4}{2\pi 0.015^2(0.15)} = \eta_{app}(-42.5)$$

$$\eta_{app} = 44 \text{ Pa.s}$$

(i) We note that the experimentally variable parameter in this rheometric study is the rotation rate of the shaft which sets the value of angular velocity at the shaft radius (a boundary condition for ω for the test fluid). In this experiment, we used an infinitely large pool of fluid bath, which may be wasteful if the amount of test fluid available is limited. Instead, we can set up a narrow annular gap containing much smaller amounts of test fluid by having a pair of concentric cylinders whereby the inner cylinder may be made to rotate. This type of flow is also known as Couette flow, and it is often adopted as the boundary conditions (at inner and outer cylinder surfaces) can be easily established and controlled and the narrow gap (much smaller length scale than the length of the cylinders) makes certain simplifying approximations possible, such as the approximation that $\dot\gamma$ is relatively constant. This greatly simplifies the mathematical steps in analysis.

Index

A

Apparent viscosity, 14–16, 18, 19, 30, 66, 67, 69, 76, 81, 86, 101, 111, 120, 146, 148, 156, 166, 173

B

Bingham fluid, 12, 14, 18, 25, 31, 34, 122–124, 167, 171–173
Breakthrough, breakthrough time, 117, 119–121, 134
Burger's model, 39–45

C

Carreau model, 75–78, 91
Complex modulus, 84, 157, 162–164
Complex viscosity, 82, 86, 88, 90, 91
Constitutive equation, 3, 14, 15, 17, 18, 23, 24, 29, 30, 33, 34, 46, 47, 56, 61, 62, 99, 109, 124, 126, 130, 131, 167
Couette flow, 136, 174
Cross model, 33, 34, 75, 78, 79, 81, 148, 149

D

Dashpot (viscous), 13, 19–22, 35–38, 40, 41, 43–47, 55, 56, 61, 83
Dilatant, 14–18, 23, 29, 30
Direction cosine, 93, 95, 107, 110

E

Elastic spring, 1, 4, 19–21, 35, 37, 38, 40, 41, 43–47, 55, 56, 61, 83, 157, 158

H

Herschel Bulkley model, 34
Hooke's Law, 4, 12, 23, 24, 35, 55, 71, 83, 84
Hysteresis, 10, 16

I

Infinite shear rate viscosity, 78
Integral strain, integral strain rate, integral strain equation, integral strain rate equation, 46, 51, 52, 54, 57, 58, 62, 64
Invariant (first and second), 94, 98

L

Laminar flow, 124, 125, 129, 130, 134
Linear viscoelastic, 55, 57, 59, 71, 82, 83, 90
Loss modulus, 82, 83, 86, 88, 91, 163, 165
Lower Newtonian region, 78, 80–82

M

Maxwell element, 19, 20, 22, 36–38, 40–42, 61, 82, 83, 85, 86
Mean residence time, 117, 119–121
Mohr's circle, 101, 103, 105, 107, 109, 111

P

Past time variable, 46, 50, 51, 57, 63, 70, 140
Phase angle, 10, 156, 159, 160
Phase shift, 10
Poiseuille flow, 136